READINGS
MARXIST SOCI(

READINGS IN
MARXIST SOCIOLOGY

*

EDITED BY

Tom Bottomore and Patrick Goode

CLARENDON PRESS · OXFORD
1983

Oxford University Press, Walton Street, Oxford OX2 6DP
London Glasgow New York Toronto
Delhi Bombay Calcutta Madras Karachi
Kuala Lumpur Singapore Hong Kong Tokyo
Nairobi Dar es Salaam Cape Town
Melbourne Auckland
and associates in
Beirut Berlin Ibadan Mexico City Nicosia

Oxford is a trade mark of Oxford University Press

Published in the United States
by Oxford University Press, New York

Introduction and selection © Tom Bottomore and Patrick Goode 1983

All rights reserved. No part of this publication may be reproduced,
stored in a retrieval system, or transmitted, in any form or by any means,
electronic, mechanical, photocopying, recording, or otherwise, without
the prior permission of Oxford University Press

This book is sold subject to the condition that it shall not, by way
of trade or otherwise, be lent, re-sold, hired out or otherwise circulated
without the publisher's prior consent in any form of binding or cover
other than that in which it is published and without a similar condition
including this condition being imposed on the subsequent purchaser

British Library Cataloguing in Publication Data

Readings in Marxist sociology.
  1. Communism and society   2. Sociology
  I. Bottomore, Tom   II. Goode, Patrick
  301   Hm51

ISBN 0-19-876108-2
ISBN 0-19-876109-0 Pbk

Library of Congress Cataloging in Publication Data
Main entry under title:
Readings in Marxist sociology.
  Bibliography: p.
  Includes index.
  1. Communism and society.   2. Marxian school of
sociology.   I. Bottomore, T. B.   II. Goode, Patrick.
HX542.R356 1983   301   83-4046
ISBN 0-19-876108-2
ISBN 0-19-876109-0 (pbk.)

Set by Macmillan India Ltd
Printed in Great Britain
at the University Press, Oxford
by Eric Buckley
Printer to the University

CONTENTS

*

Introduction	1
Note on the Texts	12

PART I SOCIOLOGY IN MARXIST THEORY

Presentation	15
KARL MARX *A Historical Science of Society*	17
KARL MARX *Society and Material Production*	20
GEORGES SOREL *Marxism and Sociological Theory*	23
MAX ADLER *The Sociology in Marxism*	30
KARL KORSCH *Marxism and Sociology*	34
GEORGES GURVITCH *Marx's Sociology*	38
Further reading	43

PART II MODES OF PRODUCTION AND SOCIAL FORMATIONS

Presentation	47
KARL MARX *Mode of Production, Civil Society, and Ideology*	49
KARL MARX *Modes of Production and Forms of Property*	51
KARL MARX *Property and Labour in Pre-Capitalist Societies*	55
MAURICE GODELIER *Tribal Society*	58
KARL KAUTSKY *The Slave-Holding System*	66
LAWRENCE KRADER *The Asiatic Mode of Production*	74
MAURICE DOBB *Feudalism*	81
Further reading	86

PART III SOCIAL CLASSES

Presentation	89
KARL MARX *The Classes of Modern Society*	92
KARL MARX AND FRIEDRICH ENGELS *The Proletariat*	93
KARL MARX *The Petty Bourgeoisie and the Peasantry in France*	95
STANISLAW OSSOWSKI *Marx's Concept of Class*	99
KARL KAUTSKY *Class, Occupation, and Status*	103

GEORG LUKÁCS *Class Consciousness* 107
KARL RENNER *The Service Class* 113
Further reading 116

PART IV THE STATE AND POLITICS

Presentation 119
KARL MARX *The Division of Labour, Civil Society, and the State* 122
KARL MARX *The State and Bureaucracy in France* 124
FRIEDRICH ENGELS *Origins and Development of the State* 125
RALPH MILIBAND *The State and the Ruling Class* 130
CLAUS OFFE *Political Power in Late Capitalist Societies* 139
RUDOLF HILFERDING *Democracy and the Working Class* 146
TOM BOTTOMORE *The Working-Class Movement* 149
Further reading 157

PART V CULTURE AND IDEOLOGY

Presentation 161
KARL MARX *Material Activity, Consciousness, and Ideology* 164
KARL MARX *The Arts and the Development of Society* 166
ANTONIO GRAMSCI *The Intellectuals* 168
LUCIEN GOLDMANN *Problems of a Sociology of the Novel* 178
WALTER BENJAMIN *The Work of Art in the Age of Mechanical Reproduction* 188
OTTO BAUER *The Nation as a Cultural Community* 194
Further reading 199

PART VI THE DEVELOPMENT OF SOCIETY

Presentation 203
KARL MARX *The Historical Tendency of Capitalist Accumulation* 206
ERIC HOBSBAWM *The Idea of Progress in Marx's Thought* 208
JÜRGEN HABERMAS *A Reconstruction of Historical Materialism* 212
MAX ADLER *Social Revolution* 218
FERNANDO HENRIQUE CARDOSO AND ENZO FALETTO *Capitalism and Dependent Development* 222

Further reading   231

PART VII MODERN CAPITALISM AND IMPERIALISM

Presentation   235
KARL MARX *The Capitalist Process of Production*   238
KARL MARX *Knowledge and the Production Process*   239
FRIEDRICH ENGELS *Crises, Joint-Stock Companies, and State Intervention*   242
RUDOLF HILFERDING *The Organized Economy*   247
KOZO UNO *The Pure Theory and the Stages Theory of Capitalism*   253
JÜRGEN HABERMAS *A Descriptive Model of Advanced Capitalism*   258
ERNEST MANDEL *Late Capitalism and Imperialism*   264
Further reading   268

PART VIII SOCIALIST SOCIETY

Presentation   271
KARL MARX *Private Property and Communism*   273
KARL MARX *Communist Society*   275
KARL MARX *Wealth in a Socialist Perspective*   278
SVETOZAR STOJANOVIĆ *Socialism and Democracy*   279
GEORGE KONRÁD AND IVAN SZELÉNYI *Intellectuals, Workers, and Political Power*   287
Further reading   292

Bibliography   293

Index   299

INTRODUCTION

*

During the past two decades the influence of Marxist thought in the social sciences, and notably in sociology, has increased dramatically. But the Marxist contribution to sociology clearly extends over a much longer period, and our purpose in this book is to show the nature and significance of that contribution, which may also be seen as the formation and development of a distinctive Marxist sociology. There has been much controversy, of course, about the relation of Marxism to sociology, and many Marxists have chosen rather to emphasize its character as political economy, a theory of history, a critical philosophy, or in some sense a completely 'new science' which is incommensurable with the existing specialized social sciences. We shall not enter into the substance of this controversy here,[1] but confine ourselves to two general observations.

In the first place, it is not a very convincing argument against Marxism as sociology to insist upon its very broad scope—the fact that Marxist ideas undeniably bear upon problems of economic analysis, historical inquiry, and epistemology or philosophy of science—for sociology too has claimed from the outset (and the claim has been sustained by its major thinkers) to be a general science of society, engaged in a study of the most diverse social phenomena, investigating both their constant and their historically variable features, making use of historical and anthropological evidence as well as the data of surveys, and constantly involved in a debate about the foundations of a scientific comprehension of social life. Even the idea of the connection between theory and practice is not unique to Marxism, although it is given a particular emphasis there and assumes specific forms.

Second, it is evident that a large number of Marxist thinkers, from the late nineteenth century to the present day, have in fact regarded Marxism as being primarily, or essentially, a system of sociology, which confronts in a critical way other bodies of sociological thought, such as those of Emile Durkheim and Max

---

[1] For a brief account of some of the major issues, see Tom Bottomore (1975), and from a different standpoint, Henri Lefebvre (1968).

Weber, and historically speaking was confronted by *them* as a rival system. Clearly, there is considerable justification for this view. In his early work, up to *The German Ideology* (1846), Marx can well be regarded as being concerned to work out a general conception of human societies as structured wholes; and in this respect he followed the same course as did other nineteenth-century thinkers who were attempting to form a rigorous concept of 'society' as the necessary starting-point of any sociological theory.

It may be claimed further that Marx was actually the first thinker to construct an adequate scientific concept of society, as a development of his conception of 'socialized humanity' (formulated in the Tenth Thesis on Feuerbach) which he himself characterized as 'the standpoint of the new materialism'.[2] But if we take this concept to be the keystone of Marx's new science of society, we must add that the originality of his theory only emerged clearly in his analysis of the specific nature of this human 'sociation' or 'social being', which produced two fundamental 'guiding conceptions', most fully expounded in *The German Ideology*, but frequently restated in later works.

The first conception is that of the historical development of human society, involving an analysis of its causes or conditions, and its general direction. In *The German Ideology* Marx defines his conception of history as resting on 'the exposition of the real process of production, starting out from the simple material production of life, and on the comprehension of the form of intercourse connected with and created by this mode of production, i.e. of civil society in its various stages as the basis of all history, and in its action as the state'; and he continues by saying that

... at each stage of history there is found a material result, a sum of productive forces, a historically created relation of individuals to nature and to one another, which is handed down to each generation from its predecessors, a mass of productive forces, capital and circumstances, which is indeed modified by the new generation, but which also prescribes for it its conditions of life and gives it a definite development, a special character.

This idea of the inescapable historicity of human social life,

---

[2] See the argument along these lines by Max Adler (1914). Partly translated in Bottomore and Goode (1978), pp. 57–62.

which pervades Marx's whole thought,[3] embraces several distinct notions. Thus on one side it presents a very broad sketch of the whole course of social development as the outcome of the human interaction with nature, or, as Marx wrote in the *Economic and Philosophical Manuscripts* (1844), 'the creation of man by human labour'.[4] This general view of development or 'progress'[5] is then made more definite and precise by means of Marx's categorization of major historical epochs or stages: 'In broad outline we can designate the Asiatic, the ancient, the feudal, and the modern bourgeois modes of production as progressive epochs in the economic formation of society' (Preface 1859).

But Marx's conception does not apply only to the broad sweep of social history. There are also stages in the development of capitalist society, some of which Marx himself outlined, and which have been the subject of a great deal of later analysis and controversy,[6] and presumably—though this is less clearly indicated in Marx's own writings—in the evolution of other forms of society as well. In recent years, particularly, Marxist historians and anthropologists have been inclined to introduce many more elements of variation and development within the broad epochs which Marx originally distinguished, and this does not seem radically inconsistent with his own approach. For Marx always recognized the importance of studying the real historical development of particular societies, in the framework of his general scheme of social evolution, as may be seen from his (and Engels's) observations on the potentialities of the village community as a basis for the transition to socialism in Russia, on the distinctive features of American society, on the prospects for a peaceful attainment of socialism in some capitalist countries with democratic political regimes, and indeed from most of his writings on the major political events of his time.

On one occasion, moreover, in his unpublished reply (1877) to

---

[3] It is expressed very clearly in the methodological introduction to the *Grundrisse* (1857) where Marx observes that in talking about production 'we are always talking about production at a definite stage of social development', and then goes on to make a careful distinction between features which are common to all epochs of production and those which constitute its development (i.e. the differentiating factors or 'determinations').

[4] See on this point the discussion by Jürgen Habermas in Part VI, below.

[5] See the comments by Eric Hobsbawm, Part VI, below.

[6] See the texts in Part VII below, and especially the excerpt from Kozo Uno.

an essay by Mikhailovsky, Marx explicitly condemned any quasi-philosophical theory which ignored real historical circumstances: 'He has to transform my sketch of the origins of capitalism in Western Europe into a historical–philosophical theory of a universal movement necessarily imposed upon all peoples, no matter what the historical circumstances in which they are placed. . . . But I must protest . . .'; and after giving a specific example Marx continues: 'The key to these phenomena can be discovered quite easily by studying each of these developments separately, but we shall never succeed in understanding them if we rely upon the *passe partout* of a historical–philosophical theory whose principal quality is that of being supra-historical'.[7]

Marx's second guiding conception, which is also clearly expressed in the passages from *The German Ideology* cited above, is concerned not with the historical development of society, but with its structure; that is to say, with the ordering and interconnection of its elements. This idea of social structure, or social system, is formulated in various ways. The most familiar, probably, is the metaphor of 'base' and 'superstructure'; in the Preface of 1859 Marx summarizes his view by saying that 'The totality of these relations of production constitutes the economic structure of society—the real foundation, on which legal and political superstructures arise and to which definite forms of social consciousness correspond.' Even in this case, however, a third term, 'social consciousness', is involved; while elsewhere Marx also employs the term 'civil society', and appears to make a distinction between the economic structure, civil society, and the state. But the distinction is not very clear; at the beginning of the Preface just cited Marx equates civil society with the 'material conditions of life', although he seems to differentiate it in some way from the economic structure by saying that its 'anatomy . . . is to be sought in political economy', and earlier, in *The German Ideology*, he wrote that 'Civil society comprises all the material intercourse of individuals within a specific stage of development of the productive forces.'[8] This poses the question of what exactly is to be understood by 'material intercourse', and in particular whether this refers simply to the 'social relations of

---

[7] For the whole passage, see Bottomore and Rubel (1963), pp. 37–8.
[8] Op. cit., vol. i, pt. i, sect. A1.

production', i.e. to the economic structure, or to some wider sphere of social interaction.

Among later Marxists, Gramsci differentiated civil society most rigorously as a separate element in social life: 'What we can do, for the moment, is to fix two major superstructural "levels": the one that can be called "civil society", that is the ensemble of organisms commonly called "private", and that of "political society" or "the state". These two levels correspond on the one hand to the function of "hegemony" which the dominant group exercises throughout society and on the other hand to that of "direct domination" or command exercised through the state and "juridical" government.'[9] More recently, structuralist Marxists, largely inspired by the work of Louis Althusser, have introduced a further distinction, in the sphere of 'consciousness', between 'ideology' and 'theory' or 'science' (conceived as a theoretical activity).[10]

Hence, in Marxist thought as a whole, taking account of all its variant forms, we are presented with a conception of social structure which comprises five elements: the economy, civil society, the state (or the political sphere), ideology, and science (or knowledge). What distinguishes Marxism as a social theory, however, is not simply this model of the major elements of society, but the attribution of some kind of primacy in the social structure to the economy (the mode of production of material life). In its most radical form this involves the assertion of a more or less total determination of the 'superstructure' by the 'base'; in short, an economic (or even technological) determinism. This has consistently given rise to difficulties in dealing adequately with the influence of other elements in social life—for example political (class) struggles, the growth of knowledge, or (to take the case made famous by Max Weber) a religious ethic—and the various attempts to restrict the comprehensive and strict determinism of

---

[9] Antonio Gramsci, *Selections from the Prison Notebooks*, ed. by Quintin Hoare and Geoffrey Nowell Smith, p. 12.

[10] This is not inconsistent with the view apparently held by Marx and Engels, though not clearly articulated or elaborated, about the partly autonomous role of science (especially in its application as technology), and even of knowledge generally, in the development of society—above all, of modern capitalist society. Marx's ideas on this subject are most fully expounded in the *Grundrisse*, though still in a fragmentary form; see particularly the excerpt on pp. 239–42 below.

the theory by introducing the idea of an 'ultimate' determination, or determination 'in the last instance', by the economy,[11] and hence the 'relative autonomy' of other elements, create fresh problems, not the least of which is to decide what exactly is meant by the 'last instance' and 'relative autonomy'. Moreover, if the relaxation of strict determinism is carried too far there is a considerable danger of lapsing into a vague notion of the pervasive 'influence' of economic phenomena, which could be accepted by anyone and would no longer sustain the originality of Marx's theory.

Some of these difficulties reappear in a new guise in the theory of social structure formulated by recent structuralist Marxists such as Althusser and Poulantzas. For if, as they claim, the key concept implicit in Marx's work is that of 'the effectivity of a structure on its elements' (i.e. a structural causality in which the interrelation of the elements is conditioned or determined by the total structure) then it seems impossible to attribute a superior causal force to any particular element (e.g. the economy), even 'in the last instance', or indeed to employ at all the conceptual scheme of 'material causality' which lay behind Marx's own analysis of the relation between base and superstructure. This kind of structuralist Marxism appears closer, in some respects, to the structuralism of Lévi-Strauss, which does not require any causally privileged social elements because it proposes to explain total social structures (which then determine their elements) by the structure of the human mind.[12]

---

[11] As Engels, in his later years, suggested in a rather *ad hoc* fashion; see especially his letters to J. Bloch (21 Sept. 1890), C. Schmidt (27 Oct. 1890), and W. Borgius (25 Jan. 1894). See also the discussion of this question on pp. 47–8 below.

[12] Maurice Godelier, in his critical discussion of Lévi-Strauss, in *Perspectives in Marxist Anthropology* (1977), pp. 44–51, seems to recognize this, and to adopt a modified structuralist position, which he describes as structural *and* materialist. Thus, while he wants to avoid a simple reductionist approach which would make the various structures in society 'epiphenomena of man's material relations with his environment' (p. 44), he also says that we must use as a 'central hypothesis' in analysing political, religious, and other elements, 'Marx's hypothesis on the determination of types of society and their evolution and modes of thought by the conditions of production and *reproduction* in material life' (p. 50). These two statements may appear contradictory, and at the least they indicate the theoretical difficulties involved in the attempt to be a Marxist *and* a structuralist.

The Marxist structuralists have usefully reasserted the importance of that part of Marx's theory which is devoted to an analysis of social structure, and employs a method (notably in the analysis of the 'commodity' in *Capital*) very close to that of modern structuralists generally, who seek to reveal, by means of theoretical constructs, the 'deep', subconscious structure of social phenomena as against the 'surface' appearance constituted by conscious intention and action. But their work raises two further problems—besides the relation between 'material' and 'structural' causality mentioned previously—which for that matter also confront, in one way or another, every kind of sociological theory; namely, the relation between structure and history, and the significance of human consciousness and intentional action in social life.

The first of these problems is clearly delineated by Piaget in his general discussion of structuralism, when he observes that a choice has to be made 'between genesis without any structure, which presupposes the atomistic interrelations made familiar by empiricism, and totalities or forms without any genesis, which always court the danger of returning to the transcendental sphere of essences, Platonic ideas, or *a priori* forms'; or else a way has to be found of transcending the opposition.[13] His own solution is to conceive a 'structured totality' as being always and simultaneously both 'structured' and 'structuring' (i.e. forming and transforming structures),[14] and so to argue for associating what he calls 'constructivism' with structuralism; while at the same time, he limits the scope of structuralism by treating it as a method—not a philosophy or world view—and insisting that it is not the only method and does not in any way exclude other approaches, especially in the social sciences.[15]

Among Marxist thinkers, Goldmann, who was greatly influenced by Piaget's ideas, expounded a synthesis of historical and structural inquiry and explanation, which he called 'genetic structuralism' and defined in the following terms: 'From this standpoint the structures which constitute human behaviour are not in reality universally given facts, but specific phenomena resulting from a past genesis and undergoing transformations

[13] Jean Piaget (1970), ch. 1.
[14] Ibid.
[15] Ibid., Conclusion.

which foreshadow a future evolution.'[16] This notion of a continuing *process* of 'structuration' and transformation, contrasted with the existence of timeless, universal, and preconstituted structures, was elaborated in sociological theory particularly by Gurvitch, from a position which can be broadly described as that of a critically revised Marxism. In his conception, both elements of Piaget's view are given a further development: first, social structure is seen as a 'permanent process, a perpetual movement of destructuration and restructuration'; and second, because social structure is conceived as only one feature of a 'total social phenomenon', which is constituted also by many other factors—mental orientations, the actions of social groups, symbols and values—it is explicitly recognized that structural analysis needs to be complemented by other types of analysis.[17]

One prominent element in the Marxist theory of society—namely, the idea of development through 'contradictions'—is examined from a structuralist standpoint in Godelier's essay on the contradictions in capitalist society.[18] Godelier identifies *two* contradictions, the first between the capitalist class and the working class, the second between the socialization of production and the private ownership of the means of production. This second contradiction is regarded as basic; and its *necessary* outcome is socialism, quite independent of conscious valuations, or more broadly, of human agency. Such a conception poses a variety of problems,[19] perhaps the most important, which has created a major division among Marxist thinkers, being that

---

[16] Lucien Goldmann (1970) p. 21.

[17] Georges Gurvitch (1962), pt. ii, ch. iv. For a fuller discussion of this approach, see Tom Bottomore (1975).

[18] Maurice Godelier (1972).

[19] Among the problems which cannot be fully discussed here is, first, that of identifying the basic contradictions in non-capitalist societies, and second, that of specifying the structure of a society without contradictions (e.g. socialism) and the factors which would then generate historical development. Marx's own view on this last question, though nowhere fully elaborated, seems to have been that socialist society would be free of contradictions because *classless*, and that the advent of such a society would mark the end of 'prehistory' and the beginning of an epoch in which human beings consciously make their own history; a view which emphasizes precisely those elements—class and consciousness—to which Godelier systematically assigns a minor or even negligible role.

which arises from the contrast Godelier establishes between the determined and necessary development of a structure on one side, and the quite secondary manifestations of this development in the conscious and value-laden actions of individuals and social groups (classes, parties, social movements of various kinds) on the other. Godelier and other Marxist structuralists seem in fact to reassert in another terminology, the doctrine of the inevitable economic 'breakdown' of capitalism which was expounded in various forms by a number of earlier Marxists[20] and frequently criticized by others.[21]

There is, evidently, a profound difference, if not an outright opposition, between the theoretical conceptions of Marxist structuralists and those of Marxist thinkers such as Gramsci, Lukács, or adherents of the Frankfurt School, all of whom, however diverse their views may be in other respects, attribute great importance in the development of society to forms of consciousness and intentional action, while at the same time they emphasize the character of Marxism as a theory of history rather than a science of structures. This fundamental disagreement— itself only one element, though a major one, in a complex of divergent views about concepts, theories and interpretations, which have regularly taken the form of distinctive Marxist 'schools'—is not, however, peculiar to Marxism. On the contrary. The disagreements within Marxism parallel, and are partly interwoven with, those which have characterized modern sociology since the end of the nineteenth century, in the repeated controversies about positivism, naturalism, historicism, the fact/value distinction, and the relation between individual or group action and impersonal 'social forces'.

In this respect too, therefore, Marxism can be seen as belonging to the mainstream of sociological thought; and the fact that there are all kinds of unresolved problems and disputed conceptions in the Marxist theory of society no more invalidates it at once and definitively as a theoretical scheme, than the existence of similar 'unsettled questions' in other sociological

---

[20] Among them Karl Kautsky and Rosa Luxemburg, although qualifications, as well as contradictory theses, are to be found in their writings on the subject.
[21] For example, by Rudolf Hilferding, who argued in *Finance Capital* and in his later writings on 'organized capitalism', that 'the collapse of capitalism will be political and social, not economic'.

theories eliminates them. Marxism has to confront these other theories, and they have to confront Marxism, in the form which it has eventually taken as a broad 'paradigm' or 'research programme'. The question then posed is whether Marxist theory provides, or opens the way for, a more plausible general account and more convincing explanations of social phenomena and events than do its rivals.

The developments of the last few decades have established Marxism in a position—though regrettably this has happened mainly in the Western world—where it does effectively challenge many alternative theoretical schemes; whereas in the past it was primarily the object of sociological criticism, was often rejected out of hand or still more often ignored, with the consequence that Marxist theoretical work was largely carried out in the context of political movements and in a relatively closed intellectual environment. From the changes that have so far taken place, involving wide-ranging controversy between diverse forms of Marxism and other schools of social thought, as well as extensive borrowings in both directions, we cannot predict with any certainty the future shape or status of the Marxist paradigm. Perhaps, in a still more fundamentally revised form, it will achieve pre-eminence; perhaps it will be incorporated into a new, radically different theoretical scheme. Some of its limitations and inadequacies are clearly indicated in the following texts; but the greatest of them, not yet sufficiently recognized, still less systematically explored, seems to us its failure to produce a theory of the nation state and nationalism.[22] At this juncture in human affairs, in the context of means of destruction unimagined by Marx or by later Marxists up to the Second World War, it may well appear that the question of capitalism versus socialism, or even the alternatives of 'socialism or barbarism' are much less significant than the issue of the survival or extinction of the human race itself, which is posed by the conflict between nation states armed with nuclear weapons. How sociological theory, Marxist or non-Marxist, can conceptualize this new condition of human society, and what can be said about the ways in which a

---

[22] Although its failure in this respect is little, if at all, greater than that of other modern sociological theories. See the comments by Tom Nairn (1977), Chapter 9, 'The Modern Janus', esp. pp. 329–31; and also Georges Haupt, Michael Lowy, and Claudie Weill (1974), and Ian Cummins (1980).

supposedly self-directing humanity might face and overcome these perils, is in the highest degree obscure; conversely, it could scarcely be clearer that any sociological theory which does not now define as a central problem the conflict between inordinately powerful states is wholly unrealistic.

NOTE ON THE TEXTS
*

Our selection of texts has been guided by the principle of making readily available substantial excerpts from a restricted number of Marxist writings, rather than attempting to cover a wider field by taking shorter passages from the very large body of material which is now available. Within these limits, however, we have tried to represent as fully as possible the principal orientations of Marxist thought over the whole period from Marx to the present day. Many of the texts have been newly translated, some for the first time.

In editing the excerpts we have occasionally omitted passages which are mainly illustrative, take up subsidiary questions, or discuss issues which would be of relatively little interest to the present day reader. These omissions are indicated by a three-point ellipsis (. . .). We have also omitted some footnotes; in particular, those which refer to other parts of the work from which the excerpt is taken. Our editorial footnotes are distinguished from those of the authors, and in some cases translators, by the sign [Eds.]. We have renumbered the footnotes to run consecutively within each excerpt.

The texts in each part of the book are preceded by a short presentation of the major problems and controversies in that field, and are followed by a brief guide to further reading which supplements the works from which excerpts have been drawn.

PART I
*
Sociology in Marxist Theory

PRESENTATION

*

The following texts illustrate the distinctive features of Marxism as a system of sociology. The two texts by Marx show that his 'new science'—formulated at a time when the term 'sociology' was scarcely known and the domain of scientific inquiry which it came eventually to designate was practically non-existent—rested upon two fundamental ideas: first, that there is a law-governed historical development of human society, and, second, that the different forms of society, and their development, are determined primarily by the relation between society and nature, by the production relation, mediated through the struggle between classes.

Later Marxist thinkers elaborated these ideas in diverse ways; sometimes, as the texts by Sorel and Korsch show, in a critical confrontation with other schools of sociological thought. It is perhaps surprising that Korsch's text, although fairly recent (first published in 1938 and partly revised in the 1950s), and claiming to examine the relationship between 'Marxist theory and modern sociological science', in fact refers only to the early period of sociology—indeed almost exclusively to Comte—and provides no analysis of the major forms of sociological thought in the twentieth century. In this respect Sorel's text, which criticizes Durkheim's sociological method, although written much earlier is considerably more 'modern'. What is illuminating in Korsch's text, however, is his indication of the way in which there emerged from the social thought of the seventeenth and eighteenth centuries—with the formulation of the concept of 'civil society', and the creation of the 'new science' of political economy, by those great 'enquirers into the social nature of man'—two antagonistic currents of sociological theory; one being that of Marx and his followers, the other that which, in the work of Durkheim and Max Weber especially, sought either to refute Marxism or to qualify some of its fundamental principles and propositions in such a way that they could be incorporated in a quite different scheme of thought.

But the critical assaults upon Marxism have always led to restatements of its basic concepts, as in the text by Max Adler, whose work as a whole provides the most comprehensive and systematic examination of the methodological foundations of a Marxist sociology; or, as the text by Gurvitch illustrates, a counter-criticism of the objections to a system of sociology founded upon Marx's ideas, including those which have been formulated by some Marxist thinkers themselves, and a fresh appreciation of the formation and the content of Marx's science of society. Today, a century after Marx's death, it is clearer than ever that his thought is one of the principal elements in the sociological tradition, no longer confined as it once was to the fringes of academic and intellectual life, or elaborated chiefly on the terrain of political movements. The questions it poses are central questions for any theory of society, its explanatory schemes major attempts to resolve them. We have shown in our introduction, in general terms, some of the difficulties which a Marxist sociology encounters, and in our presentation of specific elements of the theory in the following sections we shall examine these issues more closely.

KARL MARX

*

*A Historical Science of Society**

The fact is, therefore, that definite individuals who are productively active in a definite way enter into these definite social and political relations. Empirical observation must in each separate instance bring out empirically, and without any mystification and speculation, the connection of the social and political structure with production. The social structure and the state are continually developing out of the life process of definite individuals, but of individuals not as they appear in their own or other people's imagination, but as they really are, i.e. as they are effective, produce materially, and are active under definite material limits, presuppositions, and conditions independent of their will.

The production of ideas, of conceptions, of consciousness is at first directly interwoven with the material activity and the material intercourse of men, the language of real life. Conceiving, thinking, the mental intercourse of men appear at this stage as the direct efflux of their material behaviour. The same applies to mental production as expressed in the language of the politics, laws, morality, religion, metaphysics of a people. Men are the producers of their conceptions, ideas, etc.—real, active men, as they are conditioned by a definite development of their productive forces and of the intercourse corresponding to these, up to its furthest forms. Consciousness can never be anything else than conscious existence, and the existence of men is their actual life process. If in all ideology men and their circumstances appear upside down, as in a *camera obscura*, this phenomenon arises just as much from their historical life process as does the inversion of objects on the retina from their physical life process.

In direct contrast to German philosophy, which descends from heaven to earth, here we ascend from earth to heaven. That is to

---

* From *The German Ideology* (1846). vol. i, pt. i, sect. A. Translated by Tom Bottomore.

say, we do not set out from what men say, imagine, conceive, nor from men as narrated, thought of, imagined, conceived, in order to arrive at men in the flesh. We set out from real, active men, and on the basis of their real life process we demonstrate the ideological reflexes and echoes of this life process. The phantoms formed in the human brain are also, necessarily, sublimates of their material life process, which is empirically verifiable and bound to material premisses. Morality, religion, metaphysics, all the rest of ideology and their corresponding forms of consciousness, thus no longer retain the semblance of independence. They have no history, no development; but men, developing their material production and their material intercourse, alter, along with this, their real existence, their thinking and the products of their thinking. Life is not determined by consciousness, but consciousness by life. In the first method of approach the starting-point is consciousness taken as the living individual; in the second it is the real, living individuals themselves, as they are in actual life, and consciousness is considered solely as *their* consciousness.

This method of approach is not devoid of premisses. It starts out from the real premisses and does not abandon them for a moment. Its premisses are men, not in any fantastic isolation or abstract definition, but in their actual, empirically perceptible process of development under definite conditions. As soon as this active life process is described, history ceases to be a collection of dead facts, as it is with the empiricists (themselves still abstract), or an imagined activity of imagined subjects, as with the idealists.

Where speculation ends—in real life—there real, positive science begins; the representation of the practical activity, the practical process of development of men. Empty talk about consciousness ceases, and real knowledge has to take its place. When reality is depicted philosophy as an independent branch of activity loses its medium of existence. At best, its place can only be taken by a summing-up of the most general results, abstractions which arise from the observation of the historical development of men. Viewed apart from real history these abstractions have in themselves no value whatsoever. They can only serve to facilitate the arrangement of historical material, to indicate the sequence of its separate strata. But they by no means provide a recipe or schema, as does philosophy, for neatly trimming the epochs of history. On the contrary, our difficulties begin only when we set about the observation and the arrangement—the real depiction—of our historical material, whether of a past epoch or of the present. The removal of these difficulties is governed by premisses which it is quite impossible to state here, but which only the study of the actual life process and the activity of the individuals of each epoch will make evident.

(. . .) This conception of history, therefore, rests on the exposition of the real process of production, starting out from the simple material production of life, and on the comprehension of the form of intercourse connected with and created by this mode of production, i.e. of civil society in its various stages as the basis of all history, and also in its action as the state. From this starting-point it explains all the different theoretical productions and forms of consciousness, religion, philosophy, ethics, etc., and traces their origins and growth, by which means the matter can of course be displayed as a whole (and consequently also the reciprocal action of these various sides on one another). Unlike the idealist view of history it does not have to look for a category in each period, but remains constantly on the real ground of history; it does not explain practice from the idea but explains the formation of ideas from material practice, and accordingly comes to the conclusion that all the forms and products of consciousness can be dissolved, not by intellectual criticism, not by resolution into 'self-consciousness' or by transformation into 'apparitions', 'spectres', 'fancies', etc., but only by the practical overthrow of the actual social relations which gave rise to this idealist humbug; that not criticism but revolution is the driving force of history, as well as of religion, philosophy, and all the other types of theory. It shows that history does not end by being resolved into 'self-consciousness', as 'spirit of the spirit', but that at each stage of history there is found a material result, a sum of productive forces, a historically created relation of individuals to nature and to one another, which is handed down to each generation from its predecessors, a mass of productive forces, capital, and circumstances, which is indeed modified by the new generation but which also prescribes for it its conditions of life and gives it a definite development, a special character. It shows that circumstances make men just as much as men make circumstances.

This sum of productive forces, capital, and social forms of intercourse, which every individual and generation finds in existence as something given, is the real basis of what philosophers have conceived as 'substance' and the 'essence of man', and which they have deified or attacked. This real basis is not in the least disturbed, in its effects and influence on the development of men, by the fact that these philosophers, as 'self-consciousness' and the 'unique', revolt against it. These conditions of life, which different generations find in existence, also determine whether or not the periodically recurring revolutionary convulsion will be strong enough to overthrow the basis of the existing order. If the material elements of a total revolution—that is, on the one hand, the available productive forces, and, on the other, the formation of a revolutionary mass, which revolts not only against particular conditions of existing society but against the whole existing 'production of life', the 'total activity' on which it is based—are not present, then it is quite immaterial, as far as practical development is concerned, whether the *idea* of this revolution has been expressed a hundred times already, as is demonstrated by the history of communism.

The whole previous conception of history has either completely neglected this real basis of history or else has considered it a secondary matter without any connection with the course of history. Consequently, history has always to be written in accordance with an external standard; the real production of life appears as *a*historical, while what is historical appears as separated from ordinary life, as supraterrestrial. Thus the relation of man to nature is excluded from history, and in this way the antithesis between nature and history is established. The exponents of this conception of history have consequently only been able to see in history the political actions of princes and states, religious and all sorts of theoretical struggles, and in particular have been obliged to share in each historical epoch the *illusion of that epoch*.

KARL MARX

*

Society and Material Production*

Individuals producing in society—hence socially determined production by individuals—is naturally the starting-point. The individual and isolated hunter and fisher, with whom Smith and

* From *Grundrisse der Kritik der politischen Ökonomie* (1857–8). Introduction. Translated by Tom Bottomore.

Ricardo begin, belongs among the unimaginative conceits of the 18th century Robinsonades [stories on the model of Defoe's *Robinson Crusoe*, eds.], which in no way express a simple reaction against over-sophistication and a return to a misunderstood natural life, as cultural historians imagine. They rest upon such naturalism no more than does Rousseau's *contrat social*, which brings naturally independent subjects into relation and association by contract. This is the pretence, a merely aesthetic pretence, of the Robinsonades, great and small. It is rather the anticipation of 'civil society', which had been in preparation since the 16th century and made giant strides towards maturity in the 18th century. In this society of free competition the individual appears detached from the natural bonds, etc. which in earlier historical periods make him an appendage of a definite, limited human agglomeration. (. . .)

The farther we go back in history, the more does the individual, and hence also the producing individual, appear as dependent, as belonging to a larger whole; at first, in a still wholly natural way in the family and in the family expanded into the clan, and later in the various forms of communal society which arose out of the antagonisms and mergers of clans. In the 18th century, in 'civil society', the various forms of the social connection first confront the individual as a mere means for his private purposes, as external necessity. But the epoch which produces this point of view, that of the isolated individual, is also precisely the epoch of the most developed social relations up to now (from this point of view, general relations). Man is in the most literal sense a *zoon politikon* (a political animal), not merely a sociable animal but an animal which can acquire individuality only in society. Production by an isolated individual outside society—a rare exception, which may indeed occur when a civilized person (in whom the powers of social life are already dynamically present) is cast by accident into the wilderness—is as much of an absurdity as the development of language without individuals living *together* and talking to each other. (. . .)

Whenever we speak of production, then, what is meant is always production at a definite stage of social development—production by social individuals. It might seem, therefore, that in order to talk about production at all we must either trace the process of historical development through its different phases, or declare at the outset that we are dealing with a specific historical epoch—for example, modern bourgeois production—which is indeed our particular theme. However, all epochs of production have certain features in common, common attributes. *Production in general* is an abstraction, but a rational abstraction in so far as it really brings out and fixes the common element and thus saves repetition. But this *general* category, this common element separated out by comparison, is itself differentiated, divided into various determinations. Some determinations belong to all epochs, others are common only to a few. Some determinations will be shared by the most modern epoch and the most ancient. Production cannot be conceived without them; however, if the most developed languages have laws and attributes in common with the least developed, then that which constitutes their development must have the differentia from what is general and common. Hence the attributes of production in general must be separated out, so that beyond their unity—which arises from the fact that the subject, humanity, and the object, nature, are the same [in all epochs]—the essential divergences are not forgotten. (. . .)

It seems to be correct to begin with the real and concrete, the actual precondition; thus, to begin in economics, for example, with the population, which is the foundation and the subject of the entire social act of production. However, on closer examination this proves false. The population is an abstraction if I omit, for example, the classes of which it is composed. These classes in turn are an empty phrase if I am not familiar with the elements on which they rest, e.g. wage labour, capital, etc. These latter in turn presuppose exchange, division of labour, prices, etc. For example, capital is nothing without wage labour, without value, money, price, etc. Thus if I were to begin with the population, this would be a chaotic representation of the whole, and I should then, by means of further determinations, move analytically towards simpler concepts, from the imagined concrete towards thinner abstractions, until I had arrived at the simplest determinations. From there the journey would have to be retraced until I had finally arrived at the population again, but this time not as the chaotic representation of a whole, but as a rich totality of many determinations and relations.

GEORGES SOREL

*

Marxism and Sociological Theory*

[Durkheim's] thought is quite easy to grasp in the section where he seeks to interpret the division of labour in society. At the outset one must assume a certain differentiation produced 'simply by the growth of individual differences, implying a diversity of tastes and aptitudes'. Thus there emerged, in the first place, a phenomenon due to blind forces acting upon men by virtue of their inclinations and characters. Later, the instinct of self-preservation comes into play and drives man to undertake the effort necessary to 'maintain himself in the new conditions of existence, in which he finds himself placed as he develops historically'. This instinct has turned us in a new direction:

first, the path that he was following was found to be blocked, because the growing intensity of the struggle (for existence) made increasingly difficult the survival of those individuals who continued to devote themselves to general tasks; and on the other hand, he turned towards a more developed division of labour, that is to say, to the line of least resistance. The other possible solutions were emigration, suicide, or crime. But in most cases the bonds which attach us to our country, to life, and the sympathy which we have for our fellows, are more powerful sentiments than the habits which might divert us from a greater specialization (*The Rules of Sociological Method*, 1964, p. 93. Translated here from Sorel's citation [Eds.]).

This exposition is excellent, and I would be the first to acknowledge that it has quite a different value from the childish hypothesis of Spencer on the need for the maximum happiness or his vague declamations about adaptation. With Durkheim we are placed on the ground of real science, and we see the importance of struggle; the question is whether struggle arises in conditions quite as simple as those which the author supposes.

Socialism introduces into the study of this process a factor which sociologists systematically ignore; it does not separate the division of labour from the formation of classes. The latter, organized for struggle, have a major influence upon the division

---

* From Georges Sorel, 'Les théories de M. Durkheim', in *Le Devenir Social*, vol. i, nos. 1 and 2 (1895), pp. 23–4, 152–3, 158–63. Translated by Tom Bottomore.

of labour, by introducing forces very different from those to which Durkheim refers. Thanks to the doctrine of class struggle it is possible to trace the real historical process, whereas that which Durkheim outlines is only schematic and logical.

Thanks to the theory of classes, socialists do not attribute goals to imaginary entities, to the needs of the collective mind, or other sociological inanities, but to real men, associated in groups which are active in social life. Thus they open up a new path for psychological research, and enable it to take a prominent place in sociological investigations; they indicate the direction in which it should pursue its analyses. Psychology, assigned in this way to its proper place, provides sociology with explanatory elements, as chemistry (roughly speaking) does for the natural sciences. (. . .)

The old psychology was an obstacle to the introduction of scientific ideas about the milieu, because it was pre-eminently concerned with the unity and identity of the soul; since the 17th century it had transformed the body into a shadow of reality, an appendage. Sociology (until Karl Marx) was even less favourable to the doctrine I am discussing; its authors always had in mind theories of the political state. They investigated, for a long time, the relations between the monarch and his people; later they opposed the domain of society to that of the individual, and even today it is by no means rare to find writings in which the social problem is held to consist in finding a proper balance between these two principles. In this scheme of thought society was treated as a person, construced in accordance with the typology provided by the psychology of character.

Marx showed that none of the political, philosophical, and religious systems can be regarded as complete and self-contained with their own fundamental bases; and he demonstrated the need to posit, beneath this whole superstructure, economic relationships. What is most striking in modern society, from an economic standpoint, is the desperate struggle of interests, the anarchy of competition, and the absence of any kind of co-ordination. Whereas in former times everything seemed to be subordinated to a certain unity (more or less ideal) of the State, the new philosophy recognized the fundamental division which the earlier theories had regarded as accidental. The State was no longer seen as possessing the extraordinary prestige which it previously had; it was conceived as distinct from society, and the latter was dissolved into its relationships. From that time the notion of the milieu was created, and the whole ideology of the social took wing. It was impossible henceforth to inquire into the thought of the nation as if it were a person, since the whole of history is dominated by the class struggle. Any discussion of the nation in psychological terms gives the impression of sheer rhetoric, in which a part is confused with the whole. Thus Marx could write: 'What is collective wealth, the public fortune? It is the wealth of the bourgeoisie' (*The Poverty of Philosophy*). From the fact that the wealth of this class has increased the economists conclude that society has progressed; 'as for the working classes it is a matter of dispute whether their condition has improved as a result of the growth of this supposedly public wealth' (ibid.) (. . .)

In practice history is divided into periods of greater or lesser duration; it is accepted as an axiom that all the economic, juridical and moral phenomena of a period form a system (except for some survivals); and the attempt is made to present a clear idea of each of these systems by emphasizing some very prominent feature taken from one or other of the series of phenomena. In this way history is not denied, fairly simple classifications are obtained, and it is easy to discuss the institutions of different peoples by seeking out, from the broadest possible historical panorama, those periods which offer notable analogies with the system being considered.

But how is this theory to be interpreted and applied? As usual, Durkheim avoids dealing with the basic issue. In everyday discourse the centuries are classified by reference to their most exalted products; they are given the name of a man of genius whose works sum up the main tendencies of the age, or they are distinguished by some great political transformation, or finally, they have some ideal quality attributed to them. Taine (*Le Gouvernement révolutionnaire*, p. 129) says that 'Each society has its own elements, structure, history and context, and consequently its own vital conditions of existence. . . . In each century and each country these conditions are expressed in *norms*. In our European society the vital condition, and hence the norm, is the self-respect of every individual and his respect for others (including women and children).'

I do not deny that *norms* provide an excellent means of classifying civilizations clearly in works of popularization, but the question is whether morality is not an end product of culture; and if this is the case it is necessary, from a scientific standpoint to look for the infrastructure upon which it rests. The classification of phenomena must be based upon knowledge of the deepest levels.

This question does not only have a theoretical interest; it is of major importance for practical action. The statesman is always asking scholars for advice in order to accomplish diverse reforms, and he is given the most varied advice, based upon subjective judgements. 'In order to restore family life,' says Durkheim,

> it is not enough that everyone should see the advantages of doing so; it is necessary to set in motion the causes which alone can produce this result. In order to restore to a government its necessary authority it is essential to deal with the sources from which all authority derives; that is, to create traditions, a public spirit, etc. And to do that we must go still further back in the chain of cause and effect until we find a point at which human action can effectively intervene (*The Rules of Sociological Method*, pp. 90–1).

This advice is excellent, but extremely vague. Hence it is not surprising to find the most bizarre proposals emerging. If the best possible morality were followed in the world everything would go on perfectly and the social problem would not exist; it is somewhat naïve therefore to advise us to reform morality first of all. It has often been proposed to begin by transforming the family, and it is on this burlesque course that the *Association pour l'étude des questions sociales*, a protestant association concerned with purifying public morality, has embarked.

The standpoint of the socialist school of thought has been very precisely formulated by Engels in *The Origin of the Family, Private Property and the State*, where he says that the new forms of sexual union will not be foreseen and the problem will not be resolved until 'a new generation has grown up: a generation of men who never in their lives have known what it is to buy a woman's surrender with money or with any other social instrument of power; a generation of women who have never known what it is to give themselves to a man from any other considerations than genuine love or to refuse to give themselves to their lover for fear of the economic consequences'. Thus sexual relations are the ultimate term of those things which social reform can affect. They are also the final term of the scientific scale, and provide a means of summing up, in a comprehensive way, the tendencies of a human group. Situated as they are at the summit they do not support anything and cannot explain anything.

It is necessary to oppose to these empirical procedures of differentiation and classification a truly scientific principle based upon the theory of knowledge, and this is what non-socialist sociology has always been incapable of doing. Since all juridical relations are variable they comprise two parts, one of which can be called the *form*, the other the *matter*. In every natural science differences are founded upon a materialist definition, and this is a metaphysical principle which applies to all scientific knowledge. The *matter* of *sociology* is, in Marx's philosophy, *the system of production and exchange*, and to define our institutions is to say how this matter is (or should be) set in motion. With this principle it is possible to determine exactly what all the economic categories are at a given moment (or an average during a given period). According to Marx: 'In acquiring new forces of production men change their mode of production, and in changing the mode of production, their way of earning their living, they change all their social relations. . . . The same men who establish social relations corresponding with their material production also produce the principles, ideas, and categories which correspond with their social relations' (*The Poverty of Philosophy*).

The vulgar theory of progress then becomes useless, but at the same time it is easy to see why so much importance was attributed to this notion over a long period. When social changes could not be explained, and were even impossible to define clearly, the need was experienced to complement empirical sociology by a law which made these changes necessary. Many believed that it was man's destiny to attain happiness through work, and argued that we must resign ourselves to waiting for the day when humanity would deserve a better fate. It seemed impossible to many eminent thinkers that the march of history could be adequately explained by English utilitarianism. Reducing everything to psychological processes of an inferior sort, modelled upon those to be found in some coarse Liverpool grocer, seemed to them a monstrosity. Not wishing to appeal to miracles, to Providence, in order to fill the gap, they imagined a law of the world analogous to the 'vital force' or the forces of evolution. Durkheim indicates very well the spirit of these theories when he says that they saw 'in the milieu a means by which progress is achieved, not the cause which determines it'.

One can say the same about the doctrine of Providence and the various idealist theories. Nevertheless, the latter merit separate consideration; they should not be confused with the theological theories, and above all, not with the naïve conceptions of the simple believers in progress. Nor should one think, moreover, that idealism is a vacuous explanation, a frivolity of literary men. Socialists have often been accused of wanting to emasculate human thought and to ignore the highest creations of the human spirit. It seems to me useful to pose the question here, and so bring out more clearly the scope of the materialist conception of history.

Historical idealism is in the same situation as physiological idealism, it intervenes to enable us to co-ordinate (in a way which is more or less satisfying for the time being) phenomena which we believe to be related to some kind of determinism, but for which we have not yet discovered the true law.

It is precisely because of the complexity of the context, indeed, that we are led to formulate *artificial laws* about the relations of the part to the whole.[1] The great naturalists did not hesitate to launch bold hypotheses, which were often poorly supported, but nevertheless extremely fruitful. Claude Bernard, whose temper of mind was strongly opposed to hasty generalizations, greatly admired

> the vast horizons glimpsed by the genius of Goethe, Oken, Carus, Geoffroy Saint-Hilaire, or Darwin, in which a general conception shows us all living creatures as the expression of types which are ceaselessly transformed in the evolution of organisms and species, and in which each living creature disappears as an individual to become a reflection of the whole to which it belongs. . . . Doubtless, all these conceptions provide clarifications which guide and assist us. But to devote oneself exclusively to such hypothetical contemplation would mean turning one's back on reality.[2]

In sociology the hypotheses often issue in revolutions, and they are even more important perhaps than in physiology where they only produce therapeutic systems which are dangerous for the sick. The doctrine of the rights of man was a hypothesis, and it has certainly had some major consequences. Marx thought that in the future revolutions would assume a new character:

---

[1] Bernard (1878–9), vol. i, p. 360.
[2] *Introduction à la médecine expérimentale*, p. 159. See also Perrier (1884), pp. 134, 141, for some acute comments on the debate between Cuvier and Geoffroy Saint-Hilaire.

the social revolution of the 19th century cannot draw its poetry from the past, but only from the future. It cannot begin its own task before it has eliminated all superstition with regard to the past. Previous revolutions needed historical reminiscences in order to render themselves unconscious of their own content. The revolution of the 19th century must leave the dead to bury their dead, so as to arrive at its own content. There the phrase went beyond the content; here the content surpasses the phrase.

And he continues on the following page:

proletarian revolutions . . . criticize themselves constantly, interrupt themselves continually in their course, return to what seemed to have been achieved and begin all over again, deride with merciless thoroughness the inadequacies, weaknesses, and wretchedness of their initial efforts . . . continually recoil aghast from the indeterminate enormity of their own aims, until a situation has been created which makes any turning back impossible . . .[3]

In these passages Marx formulated in a most lucid way the future role of idealism. Previously, revolutionaries had almost always been led to adopt an illusory theory based upon myths of the past; they proceeded with an absolute conviction that they were repeating an experience confirmed by science, and these historical reminiscences prevented them from observing and discussing the facts which occurred in reality. But the great *total revolution* could not be achieved in this way, for in order to bring it about one must be convinced that it will not conform to any previous model, that it cannot be avoided, and that *the future cannot be determined*. What we can ask of social science is that it should make us aware of the development and the importance of revolutionary forces, but whereas formerly the future was grasped by means of a hypothesis accorded all the deference shown towards a scientific theory, we can now only have *indeterminate* views about the future, expressible only in the language of the artistic imagination.

Our predecessors insulated themselves from observing the present and felt certain about future solutions, but we seek to acquire a rigorous knowledge of the present and refuse to deal with the future scientifically. This change is no small achievement, and it is the consequence of the materialist theory of sociology.

[3] *The 18th Brumaire of Louis Bonaparte*, pt. i.

MAX ADLER

*

The Sociology in Marxism*

The starting-point for an understanding of Marxism is the recognition that we are not dealing here simply with an economic theory or a party-political doctrine, but that Marxism is identical with sociology. In terms of its basic theoretical outlook it is an attempt to gain knowledge of the law-governed character of social life as a whole. In this sense, it embraces every aspect of social phenomena, not only economic phenomena, which indeed it traces back to social functions and relations (and this distinguishes it from political economy). For us therefore Marxism is synonymous with sociology, which thus takes the name of its founder, just as 19th century physics was called Newtonian, or the theory of the evolution of animal species Darwinian. (. . .)

The principal characteristics of this sociological standpoint are as follows. First, the outcome of every observation of human-historical life is, in Marxism, the socialization of humanity, or rather, the *socialized human being*. That is to say, theoretical reflection on social life does not begin either with the concept of society which, even if it is not purely metaphysical, is at best an abstraction to which reality corresponds; or with the isolated individual, who is simply impossible, but with the individual as he *really* is, an individual who is always necessarily connected with other human beings in his thought and existence, since he must logically assume the existence of other beings of the same kind as himself. As soon as human beings appear they already have necessary relations of thought and action to each other. In order to live, or as Marx puts it, 'in order to produce', they enter into definite relations and connections with each other, and only within these social relations and connections does their action upon nature, their production, take place. Thus, wherever Marxism confronts the phenomena of human life—whether economic, religious, or political—it is only impelled to *theoretical* knowledge of these phenomena when it has recognized them as

---

* From Max Adler, 'Die Soziologie im Marxismus', in *Die Gesellschaft* (Sonderheft) 1925, pp. 9, 13–18. Translated by Patrick Goode. Published by permission of Verlag der Wiener Volksbuchhandlung.

social phenomena, as partial instances of this socialization of humanity, when it has traced them back, therefore, to those relations which necessarily connect the individual with his fellows. In this sense, the sociological activity of Marxism is, as it were, reductive. It does not rest until it has discovered, behind the apparently objective social arrangements and the existing social powers, which seem equally autonomous (for example, capital and the state), but also behind religion, morality, law, etc., human beings in the forms of their socialization which are specific to these phenomena. Marx's sociology brings to completion the magnificent undertaking which Ludwig Feuerbach began with the humanization of religion, but still left trapped in the abstraction of a species specific 'essence' of man.

Second, Marxism undertakes this reduction of all phenomena of social life to the underlying processes of human socialization in a particular way, by regarding *economic socialization* as the basic form of socialization, upon which all the others are dependent. This standpoint is the foundation of the materialist conception of history. (. . .)

Thus wherever Marxism confronts phenomena of social life, whether in the economy or the cultural sphere, in legal or political life, it always regards its task as being, not only to reduce these facts to the processes of socialization, but to relate the latter to the determining grounds of their economic socialization as the ultimate social motive force. The Marxist (which means for us the sociologist) has only grasped theoretically a phenomenon of political life when he has determined its place in the functional interconnections with the economic basis of the period.

Third, it follows from the reduction of social phenomena and events to that kind of human socialization, on which they are ultimately based, that socialization itself has an extremely variable and perpetually self-transforming character. Human socialization changes with experience and as a result of the activities of the human beings caught up in it. Just as external natural factors by undergoing change also change the human forms of life dependent on them, so too do men react upon, and change, their external conditions of life. Human socialization is not a passive product of external circumstances, but a continual action and reaction between the two. As Marx says: 'The coincidence of the changing of circumstances and of human activity or self-changing can only be grasped and rationally understood as revolutionary *practice*.'[1] Hence sociological observation cannot be *static*; its very object is no unchanging essence, no permanent state, but one which ceaselessly changes over time. In short, socialization is a *historical process*. Wherever Marxism confronts social phenomena, it does not only reduce them to socialization processes, and relate them to the motive forces of the economic sphere; from the outset it conceives them as historical phenomena, as circumstances which have not existed from time immemorial, which are not eternal, naturally determined conditions of human existence, but came into existence at a particular time and will again pass away. Thus for Marxism all the phenomena of social life—for example, capital, law, the state—become purely historical categories, and the same holds true for all political concepts and problems. In every case the historically developed socialization processes, in which they have acquired their concrete phenomenal form, must be investigated.

Finally, this historical process, this self-transformation of socialization, is no mere accidental and random temporal sequence, but is a law-governed process, a *process of development*. By development I mean a series of transformations in time which are not brought about by external influences, but result from a law-governed process in that which is transforming itself, and are given a determinate direction by this process. Hence every transformation series, which is a development, bears within itself its own fundamental laws. The specific law of the development of human socialization consists in the *necessary contradiction between its form and its content*, which constitutes the dialectical essence of human socialization. In its content social life is a community with others which can be dissolved; in its form it is individual existence. Thus every stage of socialization can only come into existence by the actions of individuals and is therefore immersed in the abundance of individual motives, goals, and viewpoints, which incessantly strive to make every form of socialization useful to individual interests. In this manner there arises an incessant *tension* between the form and the content of socialization, which makes every stage of socialization contradictory, and tends towards its replacement by a less contradictory form. The internal contradiction of human socialization thrusts in this way beyond every stage already attained to continually higher stages

---

[1] 'Third thesis on Feuerbach'.

of socialization. By this means, the movement of socialization in a definite direction, its development, is accomplished.

The driving force of this development is the conflict or antagonism in socialization, which at a certain level sooner or later becomes divided between different groups. These groups must ultimately become conscious representatives of the conflicting tendencies, for while one group has succeeded in exploiting the stage society has reached and strives to maintain it, the other group, for that very reason, seeks to pass beyond it. Thus the *struggle of social groups* becomes the driving force of the development of society, a struggle which at a specific economic stage assumes the form of the economic class struggle. Marx and Engels called this transformation of socialization through the inherent conflict between its form and its content, the *dialectic of the social process*. This is the defining characteristic of Marxist sociology. Wherever we have to investigate phenomena of cultural, economic or political life, we regard them not only as socialization processes which must be located in their economic context and conceived as being the outcome of historical development, but also as the product of a conflict between their social contents and their individual forms which permeates all social life. Hence we must always investigate from which group formation and group conception they proceed: whether from those which set the individual form (private interest, group or class interest) against the social content, or from the opposite; in short, what significance they have in the historical class struggle.

This class struggle itself, however, is not a sociological category, but merely a historical one. Therefore it does not belong to the essence of socialized life, but only characterizes one of its particular, though ancient, forms. It is rather the means by which socialization is led from this unsolidaristic form of its division into economically opposed classes to the solidaristic form of the resolution of economic contradictions. By contrast, the conflict between the social content of socialization and the individual forms in which it is experienced as a real sociological category continues to exist even after the abolition of class antagonisms. It simply loses its acute character, which threatens the life and development of entire social groups and remains effective only as a differentiating principle within a solidaristic communal life. This is the point at which sociology is connected with socialism, social theory with social practice. Sociology becomes the theory of socialism, socialism the praxis of sociology.

KARL KORSCH

*

Marxism and Sociology*

What is the relation between the Marxist theory and modern sociological science? If one thinks of the sociology which began with Comte and was given its name by him, then Marxism is quite alien and opposed to it. Marx and Engels disregarded both the name and the content of this sociology. Even when Marx was obliged to take notice of Comte's *Cours de philosophie positive*, thirty years after its publication, 'because the English and French make such a fuss about the fellow', he referred to 'Positivism' and 'Comtism' as something to which 'as a Party man I have a thoroughly hostile attitude', while 'as a scientific man I have a very poor opinion of it'.[1] The rejection is theoretically and historically well founded. The Marxist theory has nothing to do with the sociology of the 19th and 20th centuries as established by Comte and elaborated by Mill and Spencer.

On the contrary, it would be more accurate to conceive this 'sociology' as the opponent of modern socialism. Only from this standpoint would it be possible to understand the essential unity of all the manifold theoretical and practical tendencies (notwithstanding their great diversity) which have found expression in this discipline over the past hundred years. For later 'sociologists', up to the present day, just as for Comte after his break

---

* Originally published as ch. 1 of Karl Korsch, *Karl Marx* (London: Chapman and Hall, 1938). The present text is translated from the revised German edition of the book (Frankfurt am Main: Europäische Verlagsanstalt, 1967) pp. 3–7. Translated by Tom Bottomore. Published by permission of Europäische Verlagsanstalt.

---

[1] See Marx's letter to Engels, 7 July 1866, and to Beesly, 12 June 1871, and on Herbert Spencer, Marx to Engels, 23 May 1868. See also, Marx's ironic dismissal of 'Comtist recipes for the cookshops of the future' in his reply to the reviewer of *Capital* in the Paris *Revue positiviste*, contained in the postscript to the second edition of *Capital* (1872/3); and Engels's letter to Tönnies, 24 Jan. 1895, published in Gustav Mayer's biography of Engels, vol. ii, p. 552.

(There is a full account of Marx's relations with one of the leading English Positivists, E. S. Beesly, in Royden Harrison, 'E. S. Beesly and Karl Marx', *International Review of Social History*, vol. iv (1959) pts. 1 and 2, pp. 22–58, 208–38. Eds.)

with Saint-Simon, it has been a matter of opposing to the theory, and hence also the practice, of socialism a different form of theoretical and practical discussion of the questions which it initially posed. Marxism has a much more fundamental and direct relation to these problems which modern historical development has placed on the agenda than has all the so-called 'sociology' of Comte, Spencer, and their followers. Essentially, therefore, there is no theoretical relation at all between the Marxist theory of society and this modern bourgeois social science. Bourgeois thinkers characterize the revolutionary socialist theory of the proletariat as an 'unscientific' mixture of theory and politics. Socialists, on the other hand, dismiss the whole bourgeois social science as pure 'ideology'.

There is quite a different relation between the Marxist theory and the social inquiry of that revolutionary period of development of the English and French bourgeoisie in the 17th and 18th centuries, when indeed the name 'sociology' had not yet been invented but 'society' had already been discovered as a distinct and independent sphere of knowledge and action, and its full significance had been recognized.

According to Marx's own account in 1859,[2] he had begun to develop his materialist theory of society sixteen years earlier in a critical reassessment of Hegel's philosophy of right. At that time, in his capacity as editor of the *Rheinische Zeitung* (1842/3), he found himself 'embarrassed at first when I had to take part in discussions concerning so-called material interests'. He was thus led to take up the study of 'economic questions', and he became acquainted, though still only sketchily, with the ideas of 'French socialism and communism'. His critical discussion of Hegel led to the conclusion that 'legal relations as well as forms of the state could neither be understood by themselves, nor explained by the so-called general progress of the human mind, but that they are rooted in the material conditions of life, which are summed up by Hegel after the fashion of the English and French writers of the 18th century under the name *civil society*, and that the anatomy of civil society is to be sought in political economy'.

We see here the decisive significance which the concept of 'civil society' had acquired at that time for Marx, who was moving from Hegelian idealism to his own materialist theory. While still

---

[2] *A Contribution to the Critique of Political Economy*. Preface. (1859).

basing his penetrating materialist criticism of Hegel's idealization of the state upon the realistic statements (unexpected in an idealist philosopher) about the nature of civil society that are to be found in Hegel's own work,[3] Marx achieved through Hegel a connection with those great 'enquirers into the social nature of man'[4] who, in the preceding centuries, had established the new concept of *civil society* as a revolutionary slogan in their struggle against the old feudal economic and political system, and had already analysed, in the 'new science' of political economy, the material basis and the framework of this new bourgeois form of society.[5]

Hegel, indeed, had not derived that realistic knowledge which sets the part of his book devoted to 'civil society' in such sharp contrast with the rest of the work[6] from an independent study of the still extremely backward conditions of German society. He took both the name and the content of his 'civil society' ready-made from the French and English philosophers, political thinkers and economists. Behind Hegel, as Marx says, stand 'the English and the French of the 18th century' with their new insights into the structure and movement of society, who in turn reflect the real historical development culminating in the 'industrial revolution' in England from the mid-18th century, and in the great French Revolution from 1789 to 1815.

In constructing his new socialist and proletarian science of society Marx took as his starting-point this bourgeois theory of society, first communicated to him through Hegel. Above all, he developed bourgeois political economy (from Petty and Boisguillebert through Quesnay and Smith to Ricardo) deliberately and methodically in the sense which it had already had in the work of the great bourgeois inquirers; namely, as the anatomy of civil society. The very pungency with which Marx repeatedly pointed out that post-classical bourgeois economics (which he called 'vulgar economics') had not advanced in any

---

[3] See Marx's critical commentary on paragraphs 261–313 of Hegel's *Principles of the Philosophy of Right*, in 'Critique of Hegel's Philosophy of Right' (1843; first published in MEGA, I/1/ pp. 401–533).

[4] (In English. Eds.)

[5] See, for example, Ferguson (1767), Smith (1776).

[6] G. W. F. Hegel, *Principles of the Philosophy of Right* (1820), pt. 3, sect. 2, especially paras. 188 *et seq.* (The system of needs) and paras. 230 *et seq.* (The police).

important respect beyond Ricardo, and on many points remained inferior to his work,[7] and at the same time insisted, with regard to such phenomena as Comte's 'Positivism', that this new social-scientific synthesis was 'miserable' compared with the 'infinitely greater' achievement of Hegel,[8] shows once more the great and continuing importance which the accomplishments of that early period in the economic and social thought of the bourgeois class had for Marx's theory, even where he was able, in consequence of the new advances and goals of the proletariat as a rising, independent class, to go far beyond them. The proletariat guided by Marx's theory is therefore not only, as Engels said, 'The heir of German classical philosophy'.[9] It is also the heir of classical bourgeois economics and social inquiry. As such, it has developed further the theory inherited from the classical bourgeois thinkers, in accord with the changed historical situation.

Marx no longer considers bourgeois society from the standpoint of its first phase of development and its opposition to the feudal society of the Middle Ages. What interests him is not simply the laws of its continued existence. He treats bourgeois society in all its aspects as a historical, and hence historically transient, organization of society. He investigates the whole historical process of its genesis and development, and the inherent developing tendencies which will lead to its revolutionary overthrow. He finds two tendencies: one *objective*, in the economic basis of bourgeois society, the other *subjective*, in the new opposition between social classes arising precisely from this economic basis, not from politics, law, or morality. Thus, bourgeois society, which had previously constituted a homogeneous whole opposed only to feudalism, is now split into two opposed 'parties'. Marx conceives the supposed 'civil society' as 'bourgeois society'; that is, a society based upon class opposition in which the bourgeoisie exercises economic—hence also, political and cultural—domination over other social classes. So at

---

[7] *Theories of Surplus Value*, vol. iii (first published as 3 volumes in 4, 1905–10).
[8] Letter to Engels 7 July 1866.
[9] See the concluding sentence of Engels's *Ludwig Feuerbach and the Outcome of Classical German Philosophy* (1888). There is a similar comment, but with an additional reference to the equal importance for the Marxist theory of the 'developed economic and political conditions in England and France' in the preface to the German edition (1882) of *Socialism, Utopian and Scientific*.

last 'la classe la plus laborieuse et la plus misérable'[10] enters into the broadened perspective of social knowledge. Marx's theory recognizes the class struggle of the oppressed and exploited wage labourers in present day society as a struggle for the supersession of bourgeois society. As a materialist science of the development of bourgeois society today it is at the same time a practical guide for the proletariat in its struggle to realize a proletarian society.

The later artificial separation of sociology as a specialized discipline, which has its scientific origins in Comte, and allows the great original thinkers, who carried out the real productive work in this field in an earlier period, at most the status of 'precursors', is simply an escape from the practical, and hence also theoretical, tasks of the present historical epoch. Marx's new socialist and proletarian science, which develops further the revolutionary theory of the classical founders of social theory in a way which corresponds with the changed historical situation, is the genuine social science of our time.

GEORGES GURVITCH

*

Marx's Sociology*

The thesis which I propose to maintain here [is] that Marx's sociology—the very possibility of which is still often questioned by so-called 'official' or 'semi-official' Marxists—constitutes the basis of all his work and is an indispensable key to understanding his thought as a whole.

---

[10] This seems to be a misquotation of a reference to 'la classe la plus nombreuse et la plus pauvre' in Saint-Simon (1825). English trans. in F. H. M. Markham (ed.), *Henri Comte de Saint-Simon: Selected Writings* (1952), pp. 81–116 (eds.).

---

* From Georges Gurvitch, *La Vocation actuelle de la sociologie* (2nd enlarged edn., Paris: Presses Universitaires de France, 1963), vol. ii, pp. 220–1, 222–4, 225–8. Translated by Tom Bottomore. By permission of Presses Universitaires de France.

It is clear that Marx was not only a sociologist, but also the expounder of a political doctrine, an outstanding historian, and an economist of the first rank. It is much more difficult to regard him explicitly as a philosopher; for his implicit philosophy—which was bound up with a Promethean humanism, with the concept of 'alienation' in its multiple, often contradictory, senses, and with a call to 'transcend philosophy' by realizing it in a *praxis* guided by the knowledge of this realization—is neither very clear nor very profound. The most interesting and precise, as well as least dogmatic, element in it is the realistic and relativistic conception of the dialectic of social reality itself, and of the proper method of the science—sociology—which studies this reality as a whole. It is here that Marx integrates in a dialectical way his historical and economic studies, and appears as the first practitioner of economic sociology. (. . .)

It is true that Marx did not use the term 'sociology', and always showed an antipathy towards Comte, who invented it. In a letter of 12 June 1871 to E. S. Beesly, Marx wrote: 'My own position is totally opposed to that of Comte, and I have a very poor opinion of him as a man of science.' But Comte was only one of the founders of sociology, alongside three others: Saint-Simon, with his conception of sociology as the science of society in action; Proudhon, with his investigation of the multiplicity of groups in conflict, and of the changing balance among them, which constitute society; and finally Marx himself, with his study of the interpenetration of 'forces of production' and 'social relations of production', and the conflict between them, of which the class struggle is the most obvious manifestation, and his endeavour to penetrate the secret of the complex interaction between diverse social infrastructures and the ideological superstructures. Marx's sociology is very much closer to Saint-Simon than to Comte, whose doctrine it evidently contradicts; rejecting its static character, its neglect of social conflict, the doctrine of automatic progress, the absence of any consideration of economic sociology, and in general its 'spiritual' orientation. But Marx's hostility to Comte is the opposition of one sociological system to another; it is in no way a denial of the possibility of sociology.

In Marxist circles two arguments have been advanced against the possibility of regarding Marx as a sociologist, or even speaking of a sociology in Marx's work. The first—which is of long standing, and is still occasionally invoked (though in an attenuated form) in the USSR, in China, and in most of the People's Democracies—is that Marx, the propounder of historical materialism, which has assumed a more profound form in dialectical materialism, discovered the key to an explanation of historical processes, an explanation based upon the determination of consciousness by being, the dialectic of productive forces and relations of production, and the class struggle which is connected with it. Equipped with this key the disciples of Marx can proceed directly from the philosophy of dialectical materialism to historical explanation, guided by historical materialism. They have no need of any intermediate science, such as sociology, outside *philosophy* and *history*.

From this point of view, political economy, to which Marx devoted the last twenty-five years of his life, has no more right to exist than has sociology. *Capital* itself would be nothing more than the application of dialectical materialism to economic history. However, because the works of classical political economy are more substantial than the early studies by sociologists, Marxists of this persuasion are more inclined to attribute some value to the critical examination of bourgeois economic doctrines than to the study of current sociological theories, which are seen as a supreme example of 'petty bourgeois ideology'. (. . .)

How should one respond in defence of sociology? To say that Marx discovered a universal key for the explanation of historical processes is to deform his thought. For in all his historical writings he shows that different approaches are necessary in order to understand each particular historical conjuncture, and has recourse to a wide-ranging collaboration between the frameworks established by the sociologist and the unrepeatable elements in historical events. To look for connecting links by means of which history and sociology can collaborate and enrich each other does not in any way mean rendering sociology useless. Finally, *Capital*—which combats the dogma of abstract 'economic man', the 'fetishism' of the commodity and of capital, and the universal 'economic laws' to which the classical economists referred—can only be understood, as a scientific work, if it is regarded as an *economic sociology*, which reveals that economic phenomena, activities, and categories lose their meaning and their character when they are detached from society as a whole, from the social structure and its particular type, from the 'total social phenomenon' and the 'total human being'. From this point of view it is *entirely false to say that Marx reduced social life as a whole to economic life*. He did exactly the opposite. He revealed that economic life is only an integral part of social life, and that our conception of what takes place in economic life is falsified if we do not recognize that behind capital, the commodity, value, price, and the distribution of goods, society and the human beings who participate in it are concealed. (. . .)

The interest of Marx's recourse to the 'total human being', the 'total society', does not reside in his humanist philosophy, but in his *discovery of a new dimension*, ignored by the philosophers and the economists; social reality conceived as an ensemble of different levels, and studied by the sociologist who captures the movement of social life by the construction of types. When Marx rebelled initially against Hegel's philosophy, then against the so-called left Hegelians and Proudhon, and finally against classical political economy, it was not at all the problem of religious alienation, atheism, or humanism which was at the centre of his concerns (. . .). For Marx, his polemic against Bruno Bauer and Feuerbach, who were themselves in dispute with Hegel about the 'essence of Christianity' (the title of Feuerbach's book published in 1841), was only a pretext for breaking with Hegelianism of every kind, and at the same time for taking up a position against Saint-Simon's 'new Christianity', against the pantheistic religion of the Saint-Simonians led by Enfantin, and against the 'religion of humanity' which culminated in Comte's 'sociology'.

In fact, it was not 'religious alienation' and the problem of atheism which interested Marx, but the construction of a new science of man and of society in action, a science which we now call sociology. To be sure, Marx took the term 'alienation' from Hegel's *Phenomenology of Spirit*, but he gave it a number of sociological meanings which it had never had in Hegel. For Hegel it is first god, then his emanations, 'spirit' and 'consciousness', which alienate themselves in the world in order to bring it back to god and his living eternity. Marx, who was never properly speaking a Hegelian, could have expounded his sociology without resorting to the term 'alienation', but in so far as he did so he gave the term meanings which had never occurred to Hegel, such as *Verselbständigung* (a measure of the autonomy of the social), *Veräusserung* (externalization of the social), *Entwirklichung* (loss of reality). It is a matter here of the degree of crystallization, structuration, and organization of social reality, which may come into conflict with the elements of spontaneity in social life and fall victim to deceptive ideologies. The outcome of this process is domination and subordination which threaten both collectivities and individuals.

Sociology does not prevail only in the use of this protean term 'alienation', but in every aspect of Marx's thought; for what has been called, since Marx, 'dialectical materialism' is nothing other than a relativist and empirical sociological realism, though a relativism which is not always carried through consistently in Marx's work and of which he was not always aware. Marx's dialectic, which was not at all that of Hegel (from Plato to Leibnitz, from Leibnitz to Fichte, and from Fichte to Proudhon and Kierkegaard, one finds a series of dialectics which are very different from each other)[1], and was used by him in diverse ways in different spheres of his thought, served essentially to demolish petrified and 'unreal' concepts and to bring into prominence the movement of social totalities, within which a complex interaction takes place between forces of production, relations of production, consciousness, the creations of civilization, and ideologies. The sociology of revolutions was one of his constant preoccupations. His theory of infrastructures and superstructures reveals the tensions and the various levels of social reality, which is a major concern of any sociology. Finally, if Marx's historical studies are masterpieces of fruitful collaboration between history and sociology, they also illustrate the dialectical relation between the two disciplines, for the types of structure which the sociologist identifies are always in conflict with the underlying total social phenomena. (. . .)

We have still to answer a second objection to the possibility of a sociology which has its source in Marx; an objection which comes, like the first, from certain Marxists themselves. Thus Lucien Goldmann, in *The Hidden God*, asserts that for Marxism 'sociology is impossible, for Marxist science aims to be practical and revolutionary', and in these conditions it is impossible to separate factual judgements and value judgements. It is a matter of a 'total outlook' which acknowledges historical action but rejects 'secondary abstractions' such as sociology (p. 98). In an article of 1957, 'Y a-t-il une sociologie marxiste?'[2], Goldmann

[1] On this point see the first part of my *Dialectique et sociologie*, entitled 'Histoire des principaux types de dialectique', pp. 29–176.
[2] Subsequently reprinted in Goldmann (1959), pp. 280–302.

concludes that the Hegelian–Marxist dialectic shows that knowledge of social life 'is not science, but consciousness', and that this implies the impossibility of a Marxist sociology.

I would say in reply that if the pragmatic element in any science of man, which Marx emphasizes particularly in his *Theses on Feuerbach*, could prevent the constitution of sociology as a science, then all the other founders of sociology—Saint-Simon, Proudhon, and Comte, who all saw in it a synthesis of 'contemplation and action', a union of theory and practice, the 'return from ideas to action'—would have had to render sociology just as impossible as Marx is supposed to have done. Recognition of the fact that Marx took much more seriously, and interpreted in a much more dramatic way, than did other sociologists, the idea of *praxis*—referring to revolutionary practice on one side, to productive practice on the other, and in general to social practice as a whole—does not in the least demonstrate that 'social practice' prevents the existence of a science, sociology, which reflects precisely on this *praxis*, while attempting to 'bracket', so far as possible, its human coefficients. In his *Theses on Feuerbach* Marx was concerned above all to indicate the particularly 'committed' and ideological character of the human sciences, a character which brings them close to social and political doctrines. But he never said that the situation was hopeless, or that one could not limit substantially the pragmatic and political element by being aware of it. That is why, without waiting for the disappearance of classes and ideologies, Marx declares in the preface to *A Contribution to the Critique of Political Economy* that political economy, to the extent that it is impregnated with Marxism—that is to say, becomes an economic sociology—is a non-ideological science.

FURTHER READING

*

Avineri, Shiomo *The Social and Political Thought of Karl Marx* (1968)
Bottomore, Tom *Marxist Sociology* (1975)
Giddens, Anthony *Capitalism and Modern Social Theory* (1971), Part I.
Lefebvre, Henri *The Sociology of Marx* (1968)
Löwith, Karl *Max Weber and Karl Marx* (1982)

PART II

*

Modes of Production and Social Formations

PRESENTATION

*

The relation posited between the mode of production and the social formation[1] is a fundamental and distinctive element in Marxist theory. It has often been expressed, as by Marx himself (see text below), in terms of 'base' and 'superstructure'; but this formulation has also been severely criticized as implying a strict and one-sided economic (or even technological) determination of the whole of social and cultural life. In the 1890s, when the 'materialist conception of history' began to be more widely known and discussed, Engels responded on several occasions to criticisms of this conception, by insisting that he and Marx had never asserted more than that ' . . . the *ultimately* determining element in history is the production and reproduction of real life', that while the various elements of the superstructure exercise an influence, and interact with the economic basis, ' . . . the economic movement finally asserts itself as necessary';[2] or again, 'It is not that the economic situation is *cause, solely active*, while everything else is only passive effect. There is, rather, interaction on the basis of economic necessity, which ultimately always asserts itself.'[3]

But in spite of Engels's qualifications several problems remain. First, although the interaction of various elements in social life is recognized, the determining factor 'in the last instance'[4] is still the mode of production. Hence, the theory still asserts an

---

[1] This term has come to be widely used by Marxists to designate what Marx himself generally referred to as 'society' (as do most sociologists). In our view the terminology adopted is unimportant so long as the meaning of the concept is reasonably clear; but whatever terms are used a distinction needs to be made between: (i) human social relations, or 'sociation', as such; (ii) broad 'types' of society or social formation, e.g. tribal society, feudal society, capitalist society; and (iii) particular societies, e.g. France, Britain. For the last two classes of phenomena we may equally well employ another familiar sociological term; namely, 'social system'.
[2] Letter to J. Bloch, 21 Sept. 1890.
[3] Letter to W. Borgius, 25 Jan. 1894.
[4] Engels' term, '*in letzter Instanz*' has usually been translated as 'ultimately', or 'in the last instance', though a better translation of this last phrase would be 'in the last resort'.

economic determinism,[5] which excludes any fully independent influence of politics, culture, science, etc. (these being granted only a 'relative autonomy'); and it may be criticized, as it was by Max Weber, as constituting no more than a one-sided 'economic interpretation' which, however illuminating and fruitful it may be as a hypothesis to guide research, cannot claim a definitive scientific validity. Second, the expression 'determination in the last instance' is intolerably vague, and would only have scientific value if it were more strictly defined, not only as a general concept (being relevant perhaps only to major institutional features of society over long historical periods), but in relation to particular cases.[6]

Some recent attempts by structuralist Marxists, notably Althusser and Poulantzas, to deal with these problems, have not advanced significantly beyond the position already reached by Engels, for they still have recourse to 'determination in the last instance' and 'relative autonomy' without explicating or rendering more precise these inadequate concepts. What is most useful in Engels's own discussion is his insistence (which corresponds exactly with Marx's observations on the subject) that the 'materialist conception' is not a historical *passe-partout*, but a guide to detailed historical investigation. The following texts show how, in recent times, social scientists have used the Marxist conception of a mode of production 'determining', or (in a weaker version) 'conditioning', the general character of social relations, in empirical studies of different forms of society.

[5] And to the extent that particular weight is given, within the mode of production, to the forces of production and their development (as is often the case with Marx himself), there is also a strong element of technological determinism. It may be added that this idea of the profound, and within certain limits 'determining', influence of technology has a high degree of plausibility in the case of modern capitalism, and is indeed pervasive in non-Marxist sociology too, as is indicated by the prevalence of such concepts as 'industrial' and 'post-industrial' society.
[6] On the other hand, as we have noted in the Introduction, if the notion of 'economic determination' is qualified and weakened too much Marxist sociology loses much of its distinctiveness.

KARL MARX

*

Mode of Production, Civil Society, and Ideology*

I was led by my studies to the conclusion that legal relations as well as forms of State could neither be understood by themselves, nor explained by the so-called general progress of the human mind, but that they are rooted in the material conditions of life, which are summed up by Hegel after the fashion of the English and French writers of the eighteenth century under the name *civil society*, and that the anatomy of civil society is to be sought in political economy. The study of the latter which I had begun in Paris, I continued in Brussels where I had emigrated on account of an expulsion order issued by M. Guizot. The general conclusion at which I arrived and which, once reached, continued to serve as the guiding thread in my studies, may be formulated briefly as follows: In the social production which men carry on they enter into definite relations that are indispensable and independent of their will; these relations of production correspond to a definite stage of development of their material powers of production. The totality of these relations of production constitutes the economic structure of society—the real foundation, on which legal and political superstructures arise and to which definite forms of social consciousness correspond. The mode of production of material life determines the general character of the social, political and spiritual processes of life. It is not the consciousness of men that determines their being, but, on the contrary, their social being determines their consciousness. At a certain stage of their development, the material forces of production in society come in conflict with the existing relations of production, or—what is but a legal expression for the same thing—with the property relations within which they had been at work before. From forms of development of the forces of production these relations turn into their fetters. Then occurs a

---

* From Karl Marx, *A Contribution to the Critique of Political Economy* (1859). Preface. Translated by Tom Bottomore.

period of social revolution. With the change of the economic foundation the entire immense superstructure is more or less rapidly transformed. In considering such transformations the distinction should always be made between the material transformation of the economic conditions of production which can be determined with the precision of natural science, and the legal, political, religious, aesthetic or philosophical—in short ideological—forms in which men become conscious of this conflict and fight it out. Just as our opinion of an individual is not based on what he thinks of himself, so can we not judge of such a period of transformation by its own consciousness; on the contrary, this consciousness must rather be explained from the contradictions of material life, from the existing conflict between the social forces of production and the relations of production. No social order ever disappears before all the productive forces for which there is room in it have been developed; and new, higher relations of production never appear before the material conditions of their existence have matured in the womb of the old society. Therefore, mankind always sets itself only such problems as it can solve; since, on closer examination, it will always be found that the problem itself arises only when the material conditions necessary for its solution already exist or are at least in the process of formation. In broad outline we can designate the Asiatic, the ancient, the feudal, and the modern bourgeois modes of production as progressive epochs in the economic formation of society. The bourgeois relations of production are the last antagonistic form of the social process of production; not in the sense of individual antagonisms, but of conflict arising from conditions surrounding the life of individuals in society. At the same time the productive forces developing in the womb of bourgeois society create the material conditions for the solution of that antagonism. With this social formation, therefore, the prehistory of human society comes to an end.

KARL MARX

*Modes of Production and
Forms of Property**

The way in which men produce their means of subsistence depends in the first place on the nature of the existing means which they have to reproduce. This mode of production should not be regarded simply as the reproduction of the physical existence of individuals. It is already a definite form of activity of these individuals, a definite way of expressing their life, a definite *mode of life*. As individuals express their life, so they are. What they are, therefore, coincides with their production, with *what* they produce and with *how* they produce it. What individuals are, therefore, depends on the material conditions of their production.

This production only begins with the increase of population, and it presupposes in turn an interaction among individuals. The form of this interaction is again determined by production.

The relations between nations depend upon the extent to which each has developed its productive forces, the division of labour, and internal intercourse. This proposition is universally acknowledged. However, it is not only the relation of one nation to others, but the whole internal structure of the nation itself, which depends on the stage of development reached by its production and its internal and external intercourse. How far the productive forces of a nation are developed is shown most clearly by the degree to which the division of labour has been carried. Every new productive force, in so far as it is not a mere quantitative extension of productive forces already known (for example, bringing new land into cultivation) brings about a further development of the division of labour.

The division of labour within a nation leads at first to the separation of industrial and commercial from agricultural labour, and hence to the separation of *town* and *country*, and a clash of interests between them. Its further development leads to the separation of commercial from industrial labour. At the same time, through the division of labour, there develop further, within these different branches, various divisions among the individuals

---

* From *The German Ideology*, Sect. IA. Translated by Tom Bottomore.

co-operating in definite kinds of labour. The position of these particular groups of individuals in relation to each other is determined by the methods employed in agriculture, industry, and commerce (patriarchalism, slavery, estates, classes). The same conditions are evident, with a more developed intercourse, in the relations between nations.

The various stages of development in the division of labour are just so many different forms of property; i.e. the stage reached in the division of labour also determines the relations of individuals to one another with respect to the materials, instruments, and product of labour.

The first form of property is tribal property. It corresponds to an undeveloped stage of production in which a people lives by hunting and fishing, by cattle breeding, or at the highest stage, by agriculture. In the latter case, a large area of uncultivated land is presupposed. The division of labour is, at this stage, still very elementary, and is no more than an extension of the natural division of labour occurring within the family. The social structure, therefore, is no more than an extension of the family, with patriarchal family chiefs, below them the members of the tribe, and finally slaves. The slavery which is latent in the family only develops gradually with the increase of population and of needs, and with the extension of external intercourse, either war or trade.

The second form is the communal and state property of antiquity, which results especially from the union of several tribes into a city, either by agreement or by conquest, and which is still accompanied by slavery. Alongside communal property, personal, and later also real, private property is already beginning to develop, but as an abnormal form subordinate to communal property. It is only as a community that the citizens hold power over their labouring slaves, and on this account alone, therefore, they are bound to the form of communal property. This is the communal private property of the active citizens, who are forced to continue in this natural form of association in face of their slaves. For this reason, the whole structure of society based on communal property, and with it the power of the people, decays *pari passu* with the development of private real property. The division of labour is already more developed. We already find the opposition of town and country, later the opposition between those states which represent town interests and those which represent country interests, and within the towns themselves the opposition between industry and maritime commerce. The class relation between citizens and slaves is now completely developed.

This whole conception of history appears to be contradicted by the fact of conquest. Previously, force, war, pillage, slaughter, etc., have been postulated as the driving force of history. Here we must confine ourselves to the chief points, and therefore take only one striking example—the destruction of an old civilization by a barbarous people and the consequent formation of an entirely new social structure (Rome and the barbarians, feudalism and the Gauls, the Byzantine Empire and the Turks). For the conquering barbarians war itself is still, as already suggested above, a regular form of intercourse which is the more eagerly exploited as population increase necessitates new means of production to supersede the traditional, and for them the only possible, primitive mode of production. In Italy, however, the concentration of landed property (caused not only by indebtedness and forced sales but also by inheritance, since, as a result of loose living and the small number of marriages, the old families gradually died out and their possessions came into the hands of a few), and its conversion into grazing land (resulting from the importation of plundered and tribute corn and the consequent lack of demand for Italian corn, as well as from ordinary economic causes still operative today) had brought about the almost total disappearance of the free population. Even the slaves died out again and again and had continually to be replaced with new ones. Slavery remained the basis of the whole productive system. The plebeians, standing between the freemen and the slaves, never became more than a lumpenproletariat. Rome indeed never became more than a city; its connection with the provinces was almost exclusively political and could therefore easily be severed again by political events.

With the development of private property there appear for the first time the conditions which we shall rediscover, only on a more extensive scale, with modern private property. On the one hand there is the concentration of private property, which began very early in Rome (as the agrarian law of Licinius indicates) and developed rapidly from the time of the civil wars and especially under the emperors; on the other hand, associated with this, the transformation of the plebeian small peasantry into a proletariat, which, however, owing to its intermediate position between propertied citizens and slaves, never achieved an independent development.

The third form is feudal or estates property. If antiquity started out from the town and its little territory, the Middle Ages started out from the country. This different starting point was determined by the sparseness of the population which was scattered over a large area and which received no important increase from the conquerors. In contrast to Greece and Rome, therefore, feudal development begins in a much larger area, prepared by the Roman conquests and by the spread of agriculture associated with them. The last centuries of the declining Roman Empire and its conquest by the barbarians destroyed a number of productive forces; agriculture had declined, industry had decayed for lack of markets, trade had died out or had been violently interrupted, and the rural and urban population had diminished. These conditions and the mode of organization of the conquest determined by them gave rise, under the influence of the Teutonic military constitution, to feudal property. Like tribal and communal property it is also based on a community, but the directly producing class which confronts it is not, as in the case of the ancient community, the slaves, but the enserfed small peasantry. As soon as feudalism is fully developed the opposition to the towns reappears. The hierarchical system of landownership, and the armed bodies of retainers associated with it, gave the nobility power over the serfs. This feudal structure was, just as much as the communal property of antiquity, an association against a subject producing class, but the form of association and the relation to the direct producers were different because of the different conditions of production.

This feudal structure of landownership had its counterpart in the towns in the form of guild property, the feudal organization of trades. Here property consisted chiefly in the labour of each individual. The necessity for association against the organized robber nobility, the need for communal market-halls, in an age when the industrialist was at the same time a merchant, the growing competition of the escaped serfs swarming into the rising towns, the feudal structure of the whole country, combined to bring about the guilds. The gradually accumulated capital of individual craftsmen, and their stable numbers in an increasing population, gave rise to the relation of journeyman and apprentice, which brought into being in the towns a hierarchy similar to that in the country.

Thus, in the feudal period, the chief forms of property consisted on the one hand of landed property with serf labour chained to it, and on the other hand of individual labour with small capital commanding the labour of journeymen. The structure of both was determined by the narrow conditions of production—small-scale and primitive cultivation of the land, and handicraft industry. There was little division of labour in the heyday of feudalism. The opposition between town and country existed within each nation, and the division into estates was certainly strongly marked, but apart from the differentiation of princes, nobility, clergy, and peasants in the country, and masters, journeymen, apprentices, and soon also the rabble of casual labourers, in the towns, no division of importance took place. In agriculture it was rendered difficult by the strip system and by the emergence of the cottage industry of the peasants themselves. In industry there was no division of labour at all within the various trades and very little between them. The separation of industry and commerce already existed in the older towns; in the newer ones it only developed later when the towns entered into mutual relations.

The grouping of larger territories into feudal kingdoms was a necessity for the landed nobility as well as for the towns. The organization of the ruling class, the nobility, had everywhere, therefore, a monarch at its head.

KARL MARX

*

*Property and Labour in Pre-Capitalist Societies**

An isolated individual could no more have property in land and soil than he could speak, though he could, of course, live off it as a substance, as animals do. The relation to the earth as property is always mediated through the peaceful or violent occupation of the land by the tribe, by the community in some more or less

* From Karl Marx, *Grundrisse der Kritik der politischen Ökonomie* (1857–58), 1953 edn., pp. 385–6, 389–91. Translated by Tom Bottomore.

naturally occurring or already historically developed form. The individual can never appear here in the form of an isolated point, as a mere free worker. If the objective conditions of his labour are presupposed as belonging to him, then he himself is subjectively presupposed as member of a community through which his relation to land and soil is mediated. His relation to the objective conditions of labour is mediated through his existence as a member of the community; on the other hand, the real existence of the community is determined by the specific form of its ownership of the objective conditions of labour. Whether membership of a community appears as *communal property*, where the individual is only the possessor and there is no private property in land and soil; or whether it appears in the dual form of state and private property existing alongside each other, but in such a way that the former is a precondition of the latter, and only the citizen is and must be a private proprietor, while on the other hand his property as a citizen has at the same time a separate existence, or whether, finally, communal property appears only as a complement to private property, with the latter as its basis, while the community has no existence for itself except in the *assembly* of its members and in their association for common purposes—all these different forms of the relation of communal or tribal members to the tribal land and soil (to the earth upon which it has settled) depend partly on the natural dispositions of the tribe, partly on the economic conditions in which it exercises its ownership of the land and soil in reality, i.e. appropriates its fruits by means of labour, which in turn will depend upon the climate, the physical properties of the soil, the physically conditioned mode of exploiting it, the relation with hostile or neighbouring tribes, and the modifications introduced by migrations, historical events, etc. If the community as such is to continue in the old way, the reproduction of its members in the presupposed objective conditions is necessary. Production itself, the growth of population (which also belongs to production), in time necessarily eleminates these conditions, destroys instead of reproducing them, etc., and so the communal system declines and falls, along with the property relations on which it was based.

The Asiatic form necessarily maintains itself most tenaciously and for the longest time. This is due to the fundamental principle on which it is based; namely that the individual does not become independent of the community, that there is a self-sustaining circle of production, unity of agriculture and manufactures, etc. If the individual changes his relation to the community, he thereby changes and undermines the community and its economic premiss; on the other side, the alteration of this economic premiss, brought about by its own dialectic—impoverishment etc. In particular the influence of warfare and conquest, which, e.g. in Rome, was an essential element in the economic conditions of the community itself, destroys the real bond on which the community rests. (. . .)

It is not the *unity* of living and active human beings with the natural, inorganic conditions of their metabolism with nature, and therefore their appropriation of nature, which requires explanation, or is the result of a historical process, but rather the *separation* between these inorganic conditions of human existence and this active existence, a separation which is only fully established in the relation between wage labour and capital.

In the relation of slavery and serfdom this separation does not occur; rather, one part of society is treated by another as being itself a mere *inorganic and natural* condition of its own reproduction. The slave stands in no kind of relation to the objective conditions of his labour; rather, *labour* itself, in the form of the slave as well as that of the serf, is placed in the category of other natural beings as *an organic condition* of production alongside cattle or as an appendage of the earth. In other words, the original conditions of production appear as natural presuppositions, *natural conditions of existence of the producer*, just as his living body, even though he reproduces and develops it, is not originally established by himself but appears as the presupposition of his self; his own (corporeal) being is a natural presupposition, which he has not established. These *natural conditions of existence*, to which he relates as to his own inorganic body, themselves have a dual character: (1) of a subjective, and (2) of an objective nature. He is found as a member of a family, a clan, a tribe, etc., which then assume historically differing forms as a result of intermixture and conflict with others; and as such a member he relates to a specific nature (let us still say here, earth, land, soil) as his own inorganic being, the condition of his production and reproduction. As a natural member of the community he participates in the communal property and has a particular part of it as his possession; thus, as a Roman citizen by birth he has an ideal claim (at least) to the *ager publicus* and a real claim to a certain number of *jugera* (units) of land, etc. His *property*, i.e. the relation to the natural presuppositions of his production as belonging to him, as *his own*, is mediated by his being himself a natural member of a community. (The abstraction of a community in which the members have nothing in common but language, etc., and scarcely even that, is obviously the product of much later historical circumstances.) So far as the individual is concerned it is clear, e.g. that he relates even to language itself as *his own* only as a natural member of a human community. Language as the product of an individual is nonsense. But the same is true of property. (. . .)

The only barrier which the community can encounter in relating to the natural conditions of production—the earth—*as its own* (if we now pass on to consider settled peoples) is *another community* which already claims them as its inorganic body. *War* is therefore one of the earliest occupations of all these naturally arising communities, both for the defence of their property and for the acquisition of new property. (. . .) If human beings themselves are captured along with the land and soil as its organic accessories then they are equally captured as one of the conditions of production, and in this way slavery and serfdom arise, which soon corrupt and modify the original forms of all communities, and themselves become their basis.

MAURICE GODELIER

*

*Tribal Society**

After an unusually sustained effort to redefine and effectively use the concept of 'tribe', we arrive at a largely negative result. The classification 'tribal society' is split in two and on each side of a dividing line (the nature and origin of which remains obscure); on one side are the acephalous segmentary societies and on the other chiefdoms. Structural differences between these societies are more important than the similarities, both in quantity and

* From Maurice Godelier, *Perspectives in Marxist Anthropology*. Translated by Robert Brain (Cambridge: CUP, 1977), pp. 87–93. Reprinted by permission of Cambridge University Press.

quality, and in this sense Sahlins's 1968 attempt to regroup, in a single category, these two societal groups which he distinguished and contrasted in 1961, is a failure. This failure confirms the results of the statistical studies of Cohen and Schlegel; using Fisher's mathematical process of regressive analysis for the co-variance of multiple variables, they concluded in 1967 that 'there was no solid support for the idea of the existence of a unified social stage between the bands of hunters and gatherers and state societies'. Cohen and Schlegel also stressed that in each of these groups there are great differences in social structure and that this diversity is greatest among non-migratory agricultural and acephalous societies. This leads one to suppose that farming creates greater structural differentiation than cattle-herding or other production techniques (gathering, hunting, fishing, etc.). It is possible that detailed structural analyses of the economic systems of all these societies would reveal the existence of more than two modes of production within these two types of society, and therefore upset this over-summary classification. This does not mean that we would discover as many modes of production as there are forms of *technical* capacity in social labour, as Terray and other 'Marxists' do when they invent a mode of production each time they discover a distinct technical form of individual or co-operative labour, implying a collaboration of 2, 10 or n individuals. In fact, relations of production of similar type may involve different forms of the technical division of labour. This is a matter for future research and discussion, but in the meantime, we may conjecture the existence of several different modes of production among the different societies we call chiefdoms. Grouped under this concept, we have pastoral nomads, Turks or Mongols of Central Asia, some slash-and-burn farmers of South-East Asia, Indian hunters and fishers of the north-west coast of America, farmers from West or Central Africa and, finally, the former Chinese and Scottish 'clans', the Israelite 'tribes', etc. Perhaps what counts more in the formation of chiefdoms is less the nature of these *techniques* of production than the importance of the 'surplus' they produce. Sahlins is content to affirm that the appearance of chiefdoms and their typical economic form, linked to the practice of concentrating material wealth in the hands of the chiefs and subsequent redistribution is 'a classic example of evolutionary progress—the ability to organize greater economic and environmental diversity within a single cultural scheme, indeed within a single political group'.[1] This is a classic example of functionalist evolutionism, which takes effect for cause and cause for functional ends.

Split down the middle, the category of 'tribal' societies is hardly distinct from two other categories of society opposing each other: the 'bands' of hunter–gatherers on the one side, 'state' societies on the other. Herbert Lewis and Morton Fried have shown that both Sahlins and Service in defining segmentary tribal societies *do not distinguish* them from societies called 'bands' from which they are supposed to be distinguished. Segmentation, kinship multifunctionality, the alternation and complementarity of independent activities and forms of mutual dependence for the reproduction of their kinship relations, their ideological and political unity, perfectly characterize bands of hunter–gatherers: the Mbuti Pygmies of Zaïre (Turnbull), Bushmen of the Kalahari (Marshall), Australian aborigines (Elkin, Berndt), not to forget the Grand Basin Shoshone of North America, which Steward, toward the end of his life, gave up considering as typical examples of the simplest level of social integration, the family level.[2] It is enough to think of the internal complexity and multiplicity of the ways of life among the Murngin of Australia, the existence of patrilineal descent among the Ona of Tierra del Fuega or the Puelche and the Charrua of Patagonia, to find we are confronted by the usual collapse of empirical classifications when exceptions are more numerous than the rule. Moreover it would be wearisome to enumerate cases where the internal composition of segmentary tribes and their boundaries are as unstable as those of hunter–gatherer bands. The boundary is equally vague, though perhaps a little less so, between tribal societies 'with chiefs' and state societies. There are numerous examples of state societies in pre-colonial Africa and America, societies composed of a multitude of local 'tribes' subject to a ruling tribe, the chiefs of whom composed both the state and the dominant class.[3] Far from being radically and totally incompatible with the existence of tribal societies, the state very often only existed through the consolidating of chiefdoms and dominated tribes; sometimes created from nothing, without needing

---

[1] Sahlins (1968), p. 26.
[2] Steward in Y. Cohen (1968), vol. i, p. 81.
[3] On the Incas, cf. Murra (1956); on the Aztecs, cf. Soustelle (1955).

(like Fried or Colson) to infer from this process (verified recently by the practice of European colonial powers) that tribes and chiefdoms are exclusively *secondary* social formations, by-products of the process whereby state societies are formed.

In short, it seems that the concept of 'tribal society' covers a group of external features found in the functioning of many 'primitive' societies, the 'segmentary' character of elementary socio-economic units, the real or apparent nature of 'kinship groups' in these socio-economic units and the 'multifunctional' nature of these kinship relations. The vagueness of these criteria is such that we could apply this concept to a vast number of primitive societies juxtaposed in large congeries without clear boundaries. The most surprising thing in the history of this concept is that it has varied little in basic meaning since Lewis H. Morgan (1877). The innumerable discoveries in the field since then have only aggravated and accentuated the imprecision and difficulties without leading to any radical critique, still less to its expulsion from the field of theoretical anthropology. Content-wise, there has disappeared, through some sort of melting away, the element which was directly linked to Morgan's speculative notions, the idea of a necessary sequence of matrilineal kinship systems, followed by patrilineal ones; these notions are now out of date even for neo-evolutionists who still side with Morgan.

This shows *a contrario* that the quarrel between partisans and adversaries of an evolutionist theory of forms of social organization is being played out in a theoretical field *largely* dominated by empiricist methods; no one has ever really questioned the merits of these methods. It is not enough, like Swartz or Turner, to ignore the concept of tribe by referring no longer to it; to appeal to prudence, like Steward; or to criticize its scandalous imprecision (Neiva), its theoretical sterility and fallacies (Fried), its ideological manipulation as a tool in the hands of colonial powers (Colson, Southall, Valakazi). The evil does not spring from an isolated concept but has roots in a problem which will necessarily produce similar theoretical effects as dictated by the scientific work put into it. In Sahlins's case, the method is one of contemporary neo-evolutionist empiricism: add the limitations of one to the weakness of the other. All empiricism has a tendency to reduce the analysis of societies to a demonstration of their visible functioning traits, then regrouping the societies under various concepts, according to the presence or absence of some of these traits chosen as points of comparison. In this way we get 'abstract' concepts made up of descriptive résumés of traits abstracted from the whole to which they belong. Such concepts are neither completely empty nor useless (as Marx said apropos of concepts of 'production in general' or 'consumption in general') in that they avoid useless repetition, but they do not constitute scientific concepts.[4] They are simply legal currency for rational thought. They only become truly negative on another level when invested with an 'explanatory' value, that is, a demonstrative value, within the framework of the theoretical analysis of a precise problem, for example, the evolution of forms of society. It is at this point that neo-evolutionism adds its own impotence to the limitations of empiricism. Neo-evolutionism, in fact, utilizes abstract results, thought processes produced by the empirical operation of classing and naming societies in order to construct a hypothetical scheme of the evolution of human society. Such a scheme is not constructed from the results of analysing the *actual* evolution of societies, but is built logically from conclusions drawn from studying the evolution of nature and, in particular, the evolution of human beings. Conclusions borrowed from the natural sciences may be summed up in a few general principles: a tendency towards a greater *internal complexity* of organisms and a greater differentiation of specialized features to integrate this complexity and reproduce it. These principles have been transferred from the field of natural sciences, particularly from biology, to the field of anthropology, sociology and history, where they serve to define in advance, and in the abstract, general tendencies and the direction and principal stages of the evolution of society.

It is enough then, to select from the material furnished by anthropologists (the overwhelming majority of whom cling to empirical functionalism, and ignore the historical analysis of the societies they describe) and, without questioning the significance of this material, to choose those particular societies best able to illustrate the sociological features of a stage, a period through which humanity logically had to pass in order to arrive at the great state-type societies, beginning with small segmentary societies little differentiated from palaeolithic hunter–gatherers.

[4] Karl Marx, *Grundrisse* (1857–8). Introduction.

The chosen societies automatically become paradigms, lodged in the different divisions of a transformational scheme going from the simple to the complex, from the undifferentiated to the differentiated. In this way 'descriptive' material becomes explanatory by becoming 'illustrative'.

Let us recall the criticism such a problematique has aroused. Empiricism, in all its aspects, has been reproached for its tendency to reduce the functioning of a society to a collection of manifest or latent characteristics, then (when comparing different societies) to land in the endless dilemma of the exception and the rule. To these general criticisms we can add others particularly applicable to neo-evolutionist empiricism. This approach never seriously analyses the phenomena of reversibility, even less the phenomena of devolution, which are found in the evolution of societies; they envisage this evolution almost exclusively as a *general, one-way* movement, a march forward in general stages—apart from Julian Steward and some others who see in evolution a multilineal phenomenon. Now, many hunter–gatherer societies of South America, according to C. Lévi-Strauss, constitute 'false-archaisms', because, far from being the last representatives of the primitive stage of hunting economies in tropical forests, they are vestiges of very advanced agricultural societies, who have been pushed back by other agricultural societies from the river banks towards the forest *hinterland* where they completely lost their ability to farm. P. Clastres has revealed such a phenomenon among the Guayaki Indians,[5] and Lathrap has generalized this hypothesis for the majority of hunting societies of the tropical forest zone of America, the Tukuna, the Cashibo, Siriono, etc.[6]

Leach, for his part, has shown, apropos the Kachin of Burma, that a rank society, ruled by a chief who is, or claims to be, the last born of the direct descendants of the last of the sons of the founding ancestor in his village, can become, in certain circumstances, a society of the 'gumlao' type, without internal hierarchy and no chief, and then again a 'gumsa' chiefdom, etc. Although Leach's explanation of these reversible evolutions is not very

[5] P. Clastres, 'Ethnographie des Indiens Guayaki', *Journal de la Société des américanistes* (1968).

[6] D. W. Lathrap, 'The "Hunting" Economies of the Tropical Forest Zone of South America: An Attempt at Historical Perspective', in de Vore and Lee (1968), pp. 23–9.

convincing, since he sees in this primarily an ideological fact (the effect of successive choices by the Kachin for one or the other of two models of social organization as suggested by their system of 'values'), the analysis of such examples of reversible evolutions and even the process of devolution, is of crucial importance if one is to discover the laws of change in social structure. Jonathan Friedman was even able to prove that the social organizations of the Naga, the Wa and other peoples living near the Kachin, but with very different organizations, were so many 'involuted' forms of the Kachin system under the effect of specific economic constraints. One sees why the demonstration of the existence of such a system of transformations makes it ludicrous to class all these societies as either segmentary tribal societies or societies with chiefs.

This example shows us something more than that there is no evolution without involution; no evolution in one direction without the possibility of it in another, or several directions. It shows us primarily that there is no evolution 'in general', nor is there a 'general evolution' of mankind. Mankind is not a subject any more than are societies and their histories; history is not the story of how a germ or organism develops. To go back to a phrase of Marx, 'world history did not always exist; history as world history is a result.'[7] Finally, it is difficult to apply to past history, ideas which express recent forms of a society's evolution. The notion of 'neolithic revolution', to which the archaeologist Gordon Childe attached so much importance giving it no little panache, today seems to obscure rather than illuminate. Its implicit and modern connotation suggests rapid, deep, and brutal discontinuities, the process of animal and vegetable domestication demanding specific circumstances and several millenniums to arrive at a significant differentiation of the multiple forms (agricultural, pastoral, and mixed farming) of production assuring the subsistence of the overwhelming majority of primitive or 'tribal' societies.

To deal with facts such as these, which require a grasp of both the continuities and the ruptures, formal similarities and functional and structural differences, a method is needed which avoids reducing observed social and historical reality to increasingly fine abstractions, a method which can represent, in

---

[7] Karl Marx, *Grundrisse* (1857–8). Introduction.

thought, their internal structures and discover their laws of reproduction, non-reproduction and change. Conditions for structural reproduction change, but the changes are made according to laws expressing the particular characteristics of systems and are, therefore, constants. Our method must be capable of elucidating structures, i.e. functioning mechanisms and principles which are not directly observable. It must also be able to discover the characteristics of structural change and determine the origins and causes of such changes. To do this we must pursue our researches until we can determine the specific causality of each structure or structural level. However, this requires that we first recognize the relative autonomy of each level, exploring the connection between the form and content of the structures. If we are able to show that lineage organization constitutes the general form of social relations in two (or more) types of societies characterized by different modes of production, it is of extreme importance because it demonstrates both the relative autonomy of structural levels and emphasizes the need to go beyond a structural analysis of forms, the kind of structural morphology used by Lévi-Strauss, and utilize a structural theory of the functions and modes of articulation in social structures. The ultimate problem lies in determining the hierarchy of these functions within societies, the differential causal effects of each structure on other structures and on the reproduction of their functions and their interconnections.

If a *differential* structural causality does exist, the *decisive* problem in the comparative theory of societies, of structures as well as their histories, is to establish the cause; the determining cause in the final analysis and, therefore, the prior one in reality, even if it is not the unique or exclusive one in these structural arrangements and their transformations. From Marx to Morgan, from Morgan to Firth, from Firth to Sahlins, despite these writers' differences, the initial differential causality has been sought in the material basis of societies (neolithic revolution, industrial revolution, etc.), and in their economic organization. It is through analyses such as these that we shall be able to determine the scientific aspect of the concept of tribe, of 'tribal society', provided, of course, that we stop studying societies out of context, considering them as entities in relation to neighbouring tribes, working, as Herbert S. Lewis puts it, on *specific and limited phylogenes*. In this way, we shall gradually be able to reconstruct, not only a theory of the evolution of societies, but a theory of kinship, religion and politics within their specific structural connections and through a logic of different modes of production.

KARL KAUTSKY

*

The Slave-Holding System*

But property in land is useless without workers to till it. We have already pointed out the peculiar labour problem arising from the first formation of large landed estates. Even before the beginning of historic time, we find that the richer individuals are looking for workers who may always be counted upon, in order to add them to the household, in addition to the members of the family, who are bound to the household by ties of blood.

Such workers could not at first be had by offering them wages. To be sure, we find cases of wage labour very early, but it is always an exceptional and temporary phenomenon, for instance, help in gathering the crops. The production tools required by an independent establishment were not so extensive that a competent family could not acquire them as a rule. And family and communal ties were still so strong that any accident befalling a family and depriving it of its property could usually be counteracted by means of assistance from relatives and neighbours.

While there was but a slight supply of wage workers, there was also very little demand for them. For the household and its industry were still closely connected. If additional workers were needed for the establishment, they had to become members of the household, necessarily lacking not only a workshop of their own, but also a family life of their own, being entirely absorbed by the stranger's family. Free workers were not available under these circumstances. Even during the Middle Ages, journeymen consented to accept membership in the family of the master only as a temporary stage, as a transition to mastership and to the establishment of their own families. At this period free men could

* From Karl Kautsky, *The Foundations of Christianity: A Study of Christian Origins* (1908; English trans. New York: International Publishers, 1925), pp. 50–9. Reprinted by permission of International Publishers Co. Inc., New York.

not be permanently secured by the payment of wages as additional workers in a stranger's family. Only a compulsory detention could obtain the required additional workers for the large agricultural establishments. This purpose was served by *slavery*. Under slavery the stranger had no rights. And in view of the small size of the community in those days, the conception of the 'stranger' was all-embracing. In war, not only the captured warriors, but very often the entire population of the conquered country were enslaved and either divided among the victors or sold. But there were also means of obtaining slaves in peace time, particularly through maritime traffic, which was frequently associated with piracy in its early stages, one of the most desired booties being capable and handsome humans, who were captured on coast raids when found defenceless on the shores. In addition, the posterity of male and female slaves also passed into slavery.

The status of the slaves was at first not very bad, and they sometimes took their lot lightly. Being members of a wealthy household, often engaged in tasks contributing to comfort or luxury, they were not notably overworked. If their work was of a productive nature, it was often performed—on the big farms—with the aid of the master, and involved only production for family consumption, necessarily limited. The lot of the slave was determined by the character of his master, and by the wealth of the family to which he belonged. The masters had considerable interest in improving the status of the slaves, because it involved improvement in their own status. Besides, by constant personal contact with the master, the slave stood in a more or less human relation with the latter, and might, if he possessed wit and brightness, even become indispensable to the master, a friend as it were. Passages can be found in the ancient poets to show how free the slaves were with their masters and with what affection both sides often regarded each other. Quite frequently the slaves would be dismissed with a handsome present for faithful services and others would save enough to buy their freedom. But not a few preferred slavery to freedom, that is, they preferred life as members of a wealthy family to the lonely, meagre, and uncertain existence away from such a family.

'It must not be supposed,' says Jentsch,

> that the legal status of the slave, so repulsive to us, was taken seriously in private life and that the slave was neither considered nor treated as a

human being; up to the end of the First Punic War the lot of the slave was not a sad one. What has been said of the legal power of the head of the family over his wife and children applies also to his rights over the slaves; although legally unlimited, they were modified by religion, custom, reason, sentiment, and self-interest, and the man who was considered before the law as a commodity, subject without defense to purchase and to his master's caprice, was esteemed as a faithful fellow-worker in the fields and as a companion in the home, with whom one could chat pleasantly by the hearth after working together with him out of doors.[1]

This comradely relation was found not only on the peasant farm; even princes still did more or less work in the Heroic Age. In the Odyssey, the daughter of King Alcinous does the washing, together with her female slaves; Prince Odysseus does not challenge his rival to a duel, but to a competition in mowing and ploughing, and on his return to his homeland he finds his father working in the garden with a shovel. Besides, Odysseus and his son Telemachus are the object of the affectionate regard of their slave, the 'divine swineherd' Eumæus, who is firmly convinced that his master would have given him his liberty long ago, and also a farm and a wife, if only his master had returned.

This form of slavery was one of the mildest forms of exploitation known to us. But it changed its character when it became a means of making money, particularly when the large estates, having been separated from the household of the master, began to employ many workers.

c. Slavery in the Production of Commodities

Probably the first such properties were mines. The mining and working of minerals, particularly metallic ores, is ill-suited by its very nature for production for household use only. As soon as such industries attain even the smallest degree of development, they yield a great surplus beyond domestic needs; besides, they can attain a certain perfection only by regularly employing the labour of large bodies of workers, because the worker can in no other way acquire the necessary skill and experience, or make the necessary engineering structures profitable. Even in the Stone Age we already find great centres in which the manufacture of

---

[1] Karl Jentsch, *Drei Spaziergänge eines Laien ins klassische Altertum*, 1900. Third *Spaziergang, Der Römerstaat*, p. 237. Cf. also the Second *Spaziergang* in the same book: *Die Sklaverei bei den antiken Dichtern*.

stone implements was carried on proficiently and on a large scale, being then distributed by barter from group to group or from clan to clan. These mineral products seem to have been the first commercial commodities. They probably are the very first to have been produced with the intention of serving for barter.

As soon as a mining operation had developed over a deposit of valuable minerals, and had passed beyond the limits of the most primitive surface mining, it required larger and larger bodies of workers. The need for such workers might easily exceed the number of free workers that could be recruited from the ranks of the clan owning the mine. Wage labour could not permanently supply numerous bands of workers; only compulsory labour by slaves or condemned criminals could assure the necessary number of workers.

But these slaves were no longer producing only utensils for the limited personal requirements of their master; they worked so that he might make money. They were not working for his consumption of sulphur, iron or copper, gold or silver, in his own household, but for his sale of the mined products, to put him in possession of money, that commodity that can purchase everything, all enjoyments, all power, and of which one can never have too much. As much labour as possible was ground out of the workers in the mines, for, the more they worked, the more money their owner made. And they were fed and clothed as poorly as possible, for their food and clothing had to be *bought*, had to be paid for in *money*; the slaves in the mine could not produce them. While the owner of a wealthy agricultural establishment could do nothing else with his surplus of articles for consumption than lavish them on his slaves and guest-friends, the case with commodity production was different; the less the slaves consumed, the greater was the gain in money from the industry. Their situation became worse and worse as the industry became larger, thus removing them more and more from the master's household, housing them in special barracks whose dismal bareness contrasted sharply with the luxury of the former household. Furthermore, all personal contact between master and slave ceased, not only because the workshop was now separated from his household, but also because of the great number of workers. Thus it is reported in Athens at the time of the Peloponnesian War that Hipponikos had six hundred slaves working in the Thracian mines and Nikias one thousand. The slave's position now became a terrible scourge for him; while the free wage worker might after all make a certain selection among his masters and might at least under certain favourable circumstances exercise a certain pressure on his master by refusing to work, and thus resist the worst encroachments, the slave who ran away from his master or refused to work for him might be slain on sight.

There was only one reason for sparing the slave, the reason for which one spares cattle: the cost of buying a new one. The wage worker costs nothing, and if the work destroys him another will take his place, but the slave had to be bought; if he died before his time, his master was the loser. But this reason had less and less influence when slaves were cheap, and there were times when the price of a slave was extremely low, when constant foreign and domestic wars threw numerous captives on the market.

Thus in the third war of the Romans against Macedonia, seventy cities were plundered in Epirus, in the year 169 BC, *on a single day*, 150,000 of their inhabitants being sold as slaves.

According to Böckh, the usual price of a slave in Athens was 100–200 drachmas ($20–$40). Xenophon reports that the price varied between fifty and a thousand drachmas. Appianus says that in the Pontus on one occasion prisoners of war were knocked down at four drachmas (a trifle over 75 cents) each. When Joseph's brothers sold him to Egypt he brought in only 20 shekels ($4.50).[2]

A good riding-horse was much more expensive than a slave, as its price at the time of Aristophanes was about 12 *minae*, or almost $250.

But the very wars which furnished cheap slaves also ruined many peasants, since the peasant militia then constituted the kernel of the armies. While the peasant was waging war his farm would go to pieces for lack of workers. The ruined peasants had no other resource than to take to banditry, unless they had the opportunity to go to a neighbouring city and eke out their livelihood as artisans or as part of the *lumpenproletariat*. Many crimes and criminals were thus produced that had not been known in earlier days, and the pursuit of these criminals furnished new slaves, for jails were as yet unknown, being a

---

[2] Herzfeld, *Handelsgeschichte der Juden des Altertums*, 1894, p. 193.

product of the capitalist mode of production. Persons not crucified were condemned to compulsory labour.

Over certain periods there were therefore available extremely cheap hosts of slaves whose status was very wretched. The Spanish silver mines, among the most productive of antiquity, are an excellent illustration. 'At first,' Diodorus says of these mines,

> ordinary private individuals undertook the mining and gained great wealth thereby, since the silver ore was not deep in the ground and was present in great abundance. Later, when the Romans had become masters of Iberia (Spain), a large number of Italians were attracted to the mines, gaining great wealth through their avarice, for they bought a number of slaves and handed them over to the mine supervisor. . . . The slaves who have to work in these mines make incredible sums for their masters; but many of them, working far below the ground, exerting their bodies day and night in the shafts, die from overwork. For they have no recreation or recess in their work, but are driven on by the whips of their supervisors, to bear the worst discomforts and work themselves to death. A few who possess sufficient physical strength and a patient equanimity, are able to bear this treatment, but this only prolongs their misery, the immensity of which makes death appear more desirable to them than life.[3]

While patriarchal domestic slavery is perhaps the mildest form of exploitation, slavery in the service of greed is surely the most abominable. The technical methods of mining under the given circumstances made it necessary to employ a large-scale production with slaves, in the mines. But in the course of time a demand arose for the production of commodities on a large scale by slaves in other branches of industry. There were communities that were far superior to their neighbours in military power, and these found war so profitable that they never tired of it. Warfare furnished an inexhaustible supply of new slaves which it was sought to put to profitable work. But these communities were always connected with great cities. When such a city, because of its favourable situation, became a great trading place, commerce alone would attract many persons, and if the city was generous in its grants of citizenship to strangers, it soon became richer in

---

[3] Diodorus Siculus, *Historische Bibliothek*, vol. xxxvi, 38. Compare the quotation from the same work, vol. iii, 38, on the Egyptian gold mines, to which Marx refers in his *Capital*, vol. i, chap. 8, 2, note 43.

population, and also in means, than the other neighbouring communities which it subjected. Plundering and exploiting the surrounding country was a further source of increasing wealth for the city and its inhabitants. Such wealth would stimulate the need for great building operations, either of a hygienic nature—sewers, aqueducts; or of aesthetic and religious nature—temples and theatres; or of military nature—encircling walls. Such structures could at that time be best produced by great masses of slaves. Contractors arose, who bought great numbers of slaves and executed various constructions for the state with their labour. The large city also furnished an extensive market for great masses of foodstuffs. With the low price of slaves, the most extensive surplus was produced by agricultural establishments working on a large scale. To be sure, the technical superiority of large-scale production in agriculture was at that time by no means an accomplished fact. In fact, slavery was less productive than the labour of free peasants, but the slave, since his labour power did not need to be spared and he could be driven to death without regret, produced a greater *surplus* over and above the cost of his maintenance than did the peasant, who had then not yet learned to appreciate the blessings of overwork and was accustomed to a life of ease. In addition, slave labour had the advantage, precisely in such communities, that the slave was freed from military service, while the peasant might at any moment be taken from the plough by the duty of defending his country. Thus, in the economic territory of such large and warlike cities, large-scale agricultural production by slaves began. It was brought to a high level by the Carthaginians; the Romans became acquainted with it in the wars with Carthage, and when they annexed large territories from their great rival, they also annexed the practice of large-scale agricultural production, which they further developed and expanded.

Finally, in large cities where there were many slaves practising the same trade, and also a good market for their products, it was a simple matter to buy up a large number of such slaves and put them to work in a common factory, so that they might produce for the market as wage workers do today. But such slave manufactures attained importance only in the Hellenic world, not in the Roman. Everywhere, however, a special kind of slave industry developed together with large-scale agricultural production, regardless of whether such production was a mere plantation furnishing a certain species such as grain by factory methods for the market, or whether it chiefly served the home consumption by the family, by the household, and therefore had to furnish the very varied products which the latter required.

Agricultural work is peculiar in that it demands a large number of workers only at certain seasons of the year, while at other seasons—particularly in winter—it requires but a few. This is a problem even for modern large-scale agricultural establishments; it was a harder problem under the system of slave labour. For the wage worker can be dismissed when not needed and reemployed when needed. How he gets along in the interval is his own business. On the other hand, the large-scale farmer could not sell his slaves every autumn and buy new ones in the spring. He would have found such a practice very expensive, for in the fall they would have been worth nothing and in the spring a great deal. He therefore was obliged to try to keep them busy during periods in which there was no farming. The traditions of a combined agriculture and industry were still strong, and the farmer still worked his own flax, wool, leather, timber, and other products of his land into clothes and implements. Therefore the slaves of large-scale agricultural enterprises were employed, during the time when farming was idle, at industrial tasks such as weaving, and the manufacture and working of leather, the making of wagons and ploughs, the production of pottery of all kinds. But, when the production of such commodities had advanced to a high level, they manufactured not only for their own establishment and household, but also for the market.

When slaves were cheap, their industrial products could also be made cheap, as the latter required no outlay in money. The estates, the latifundia, furnished the food and raw materials for the workers, and for the most part even the tools. And as the slaves had to be kept alive in any case during the period in which they were not needed for farming work, all the industrial products which they produced beyond the needs of their own establishment and household constituted a surplus that might allow a profit even at low prices.

It is not to be marvelled at that a free and healthy artisan class could not develop in the face of this competition from slave labour. In the ancient world, particularly the Roman world, the artisans remained wretched fellows, working mostly alone without apprentices, and usually in the customer's house, with materials furnished by the latter. A healthy artisan class, such as later developed in the Middle Ages, is entirely absent. The guilds remained weak, the artisans constantly at the mercy of their customers, most of whom were large landed proprietors, as whose clients they often led a very parasitic existence on the threshold of the *lumpenproletariat*.

But large-scale production with slave labour was only powerful enough to prevent a healthy growth of free industry and a development of its technique, which always remained at a low level in ancient times, as was natural in view of the artisan's poverty; but the artisan's skill might on occasion become highly developed, his tools remaining wretched and primitive. But the case in the large-scale enterprises was no different; here also slavery had the same inhibitive effect on all technical development.

LAWRENCE KRADER
*
The Asiatic Mode of Production*

The countries of Asia which provided the historical data for the theory of the Asiatic mode of production were India in the first place, also China and Persia. These are lands in which agriculture has been the predominant basis of subsistence over thousands of years, which has been reported by written history during the entire period. The great majority of the people of these countries lived in villages, forming communities in direct relation with the soil, cultivating it by means which had changed but little from prehistoric times; but at last the continuity was broken up by the colonialist incursions, at the time that Marx engaged in his researches into the subject; the dual means of the incursions at the hands of the colonialist powers of Europe were trade and armed force. European capitalism had means superior to the Asiatic mode of production in both respects. [. . .]

---

* From Lawrence Krader, *The Asiatic Mode of Production* (Assen: Van Gorcum & Comp. B.V., 1975), pp. 286, 290–6. Reprinted by permission of Dr Lawrence Krader.

The agriculture in the Asiatic mode of production was dependent on the storage and conducting, retention, damming, coffering or deflection of the water courses, generally of their management and control. Much of India, Southeast Asia, and China lies within the zone of the monsoon rains, which are seasonal downpourings, providing great quantities of water in one season of the year and dearth in another. The water must be collected in the season of plenty and released in the season of want. For this reason, a great economic and social value was placed on the development of the sciences for the predictions of the water supplies and their control, and the technology necessary for the storage and the distribution of the water. The early agriculture was developed in the valleys on the Nile, Tigris, Euphrates, Ganges, Yellow, and Yangtze Rivers, valleys which were inundated by the overflows of the river banks during the seasons of abundance of water, and provided the riches for the fertilization of the soil. The flooding of the rivers was the subject of the predictions by the priestly astronomers in the ancient days.

In India the Brahmins who made the predictions of the seasonal supplies of water were the residents of the villages. On the other hand, in ancient Egypt, the science and technology relative to the control of the water lay in the hands of the priestly astronomers who were direct agencies of the State. The labour of ditch-digging, and dam-building for the irrigation systems was provided by the agricultural producers. Superficial differences in the theory of the Asiatic mode of production have been introduced by the consideration of the different models based upon the different lands which are included in the Asiatic mode of production as a category, whether India, China, Egypt, Ceylon, Persia, Burma, or other countries mentioned in this connection. The centralization of the management of water control is by no means a feature common to all; it is not a specific and determining characteristic of the Asiatic mode of production.

There is a widely held opinion that the Asiatic mode of production and the Oriental despotism are connected social phenomena. The despotism in the Orient is sometimes accounted for by the fiction that the sovereign was the owner of all the land in his realm, sometimes by the fiction that the State was the sole, centralized power whereby the water was controlled and managed, the water control being a necessary condition for agriculture in the Asiatic mode of production. That the control of the water supply is indeed a necessary condition for the agriculture in the Asiatic mode of production may be taken as given, but the agency of the State in this connection does not take the form that has been attributed to it by the proponents of this branch of the theory. The role of the State through its agencies of water control is attested in some societies associated with the category of the Asiatic mode of production, but not in others. From this it follows that while despotism is found in the various societies in this category, it is not founded upon its centralization of the control over the water supply. The coincidence of political despotism and water control is an extrinsic, and not an inherent characteristic of the relations of the Asiatic mode of production. Such despotic rule as has been noticed in the traditional history of India was not founded *in principle* on the centralization and management of the water supply.

The foundation of the despotism is to be sought elsewhere: just as there were no alternatives to village employment, which made the traditional occupations stable, so there were no alternatives to the sovereign power, which made that power absolute. The dynastic struggles merely altered the succession without altering the mode of rulership.

By raising the question of the processes and relations of the Asiatic mode of production we raise at the same time the question of the social evolution of mankind.

The village community of the Asiatic mode of production, according to the theory, was the location of the transition of mankind, the bearer of the transition, from the undivided society to the society divided into classes and opposed within itself. The category of the village community in this connection points to societies of the Asiatic type as the earliest form of class-divided societies, or the prototype for all such societies, whether the earliest chronologically or not. On the other hand, the theory of the gens was advanced, according to which another social institution and another type of society was made into the bearer of the transition, and served as its model. The prototype, the location, again setting aside the question of priority in time, was now transplanted to ancient Greece and Rome. But the gens and the village community differ not only in the geographic location of the transition; each points in an opposed direction, and a different problem of social evolution is posed by each. The village community develops into its given historical form in the process of the formation of opposed social classes and the State; the gens by its collapse and dismemberment gives way to the new form of society. The village community arches over the transition from primitive to civilized society, and new relations of production as well as new social forms corresponding to them are thereby released. The gens points backward, and only by its having been overcome, by its elimination and sublation, are the new production relations and the new social forms developed. The two categories stand in a dialectical opposition to each other.

The village community within the Asiatic mode of production contained and set in motion new productive forces, which transformed the history of mankind. In its internal evolution it developed commodity exchange between the communities, and thereby the production of commodities was generated; it created surplus production, surplus value extracted by the agencies of the State in the form of involuntary drafts of labour. Out of the village community was generated the social division of labour thereby. In the evolution of the village community, its productive forces were transformed, the division of the social classes and the formation of the political society established. We draw the attention of the reader to this point of evolution, which is the indicator of the changeover; the generator of the change was the release of the productive forces by the increase in the density of the interrelations of the villages, and in particular, of commodity exchange. The transformation of the society of the Asiatic mode of production first took place on the basis of the exchange relations between communities, and later of the relations of production, just as, at the end of its history, the transformation of the Asiatic mode of production was effected by the introduction into the social whole of European commerce. The change was effected in both cases by relations to the outside, in the former case on the scale of the community, in the latter on the scale of the society. In the latter case it was also imposed, in part, from without. The increased density of the exchange relations in the early period of the history of the Asiatic mode of production had as its effect the production of a surplus which was extracted by the agencies of the State. This was not a balanced relation. The generative force lay in the exchange relations, and not in the production relations. The agency of the State did not increase the productivity of agriculture.

Much has been written about the stagnation of the Asiatic mode of production; to be sure, these societies showed a backward form to the outside observer. Nevertheless, the village communities in the Asiatic mode of production carried out the transformation of human society and stood to the primitive production in the same way that the capitalist mode of production stands to the Asiatic. Each constitutes an evolutionary movement, opening out productive forces and social moments relative to its forerunners.

Within the Asiatic mode of production, both private property and capital formation were weakly developed; the observations of the Europeans who wrote of the weakness of private property were just. But they feared that the weak development of property in the form that it had taken in Europe at that time was the determinant of the despotic power in the Orient; the fears of these seventeenth- and eighteenth-century men were without foundation. The Oriental despotism did not result from the weakness of private property; to attribute the unchecked and arbitrary power of the sovereignty in the Orient to the weakness of private property is an argument of economic determinism, but of a crude sort, for it reduces the economic relations to their formal side. (...)

In the capitalist mode of production, ownership of the means of production is concentrated in private hands, and ownership of the same by institutions of the State or by the community is a residual category, being found where private property has been given up, or where it has not been moved in. In the Asiatic mode of production, these relations are turned around: ownership of the means of production, of the land in the first place, here rested with the community, or with the overarching community, the State; in this case, private ownership of the means of production becomes the residual category. The two developments, capitalism and the Asiatic mode of production, have different kinds of property relations. (...)

The relation between the Asiatic and the capitalist modes of production is indeed a dialectic. The capitalist stands to the Asiatic mode of production as a development out of it and as its antagonist. The capitalist practices were imposed by force in the lands of the Asiatic mode of production, which became their colonies. In Asia during the seventeenth-nineteenth centuries, capitalism arose, at least in a modest degree, but the forces that led to this were external to a major extent, and internal to a minor extent. It is this dialectic that is muddied and obscured by comparisons of weak and strong property relations and forms.

Further: the relations of the Asiatic and capitalist modes of production have a bearing on particular historical courses, the conquests in Asia by European powers in the capitalist period. They have at the same time a bearing on development of political economy and society on a worldscale. The Asiatic mode of production both precedes and gives rise to the capitalist in a general way. The relations of capital and labour, as well as forms of property, are related as developments and antagonisms between the two, they are related directly and mediately, in theory and in practice. Directly and practically they are related by colonial conquest. On the other hand, they are related mediately; here we take into account the other modes of production, such as the classical and feudal. Capitalism, from this point of view, is a realization of potentialities in political economy and society, in relations of labour to the means of production, of capital formation, of commodity exchange and production, class relations, surplus production and extraction, and the evolution of the State. These potentialities already exist in the Asiatic mode of production, and are realized in the extent that we have seen. The capitalist and Asiatic modes of production are related in theory, this relation has a bearing on the theory of the social evolution of mankind.

The application of criteria of private property and despotism to Asia by mercantilists and others, even down to the present day, decked the societies that they found in borrowed clothes. Moreover, the relation drawn between private property and despotism is a superficial one. The despotism did not rest on the property form. On the contrary, both are the result of other relations, within the communities, between the communities, and between the communities and the agencies of the State. Thus, the strength of the age-old communities was an important factor in determining the forms of property ownership. The weak development of commodity relations between these communities was related in turn to the self-sustaining strength of the village communities. Thus, that very strength was a condition of the weakness, or at any rate, the slow development, of the Asiatic mode of production.

Community and society. Human beings are social creatures,

living only in society; the human being is a network of social relations. The primitive social life of mankind, however, is life in a community. The break away from the community fully begins only with the development of political society. The society of the Asiatic mode of production is near the beginning of this development; it is a transitional one in the sense that the political society has been developed, but, at least in the villages, the social and community relations do not yet diverge, the community life is the social life. The division of social labour in these communities is the village division of labour.

These community relations are in part kinship relations, but to reduce the entirety of the community relations, hence of the social relations, to those of kinship, whether by descent or by marriage, is a gross reduction and a simplification. L. H. Morgan reduced the question of the community to that of the consanguineal unity, the gentes, bound as they are, in his conception to ties of descent; this is likewise the error of H. S. Maine, who thought that the ancient Indian village community was in its origin the Joint Family. The social bonds of the ancient community were not only those of kinship; these communities, whose traces are found in ancient Roman as well as ancient Oriental societies in the historical periods, were at once social groups deriving their livelihood from their land. They worked upon the land because they were related as kin, as neighbours, communally and hence socially; at the same time, because they worked upon it their relations as kin were socially acknowledged; because they worked the land, its passage from one generation to the next in the same descent line was accepted by the social unity as a whole, in this case, the community. Social labour was divided in the ancient Indian community, or the labour in the ancient Indian community was labour in society, which was divided.

The great period of social evolutionary theory was the nineteenth century in Europe, when the progress of mankind was evidently proved. The weakness of the theory at that time was its simplism, its naive progressism, its advocacy of a grand teleology which, it was proposed, aimed at the establishment of the contemporary social state of the European model of one sort or another. The theory was uncritical, save for vague averments by L. H. Morgan. It has had few developments as such in the twentieth century, whether in critical or uncritical form, because of its many weaknesses, while writers on the subject have not coupled their theorizings with social criticism. On the contrary, the theory of the Asiatic mode of production contains the theory of transition from the primitive to the civilized condition on the one hand, and the critique of the latter on the other. It is the civilized and their *mission civilisatrice* who by colonialist practice imposed European power in many parts of the world, and brought the Asiatic mode of production to an end.

MAURICE DOBB

*

*Feudalism**

This country has not been immune to discussion about the meaning of Feudalism, and usages of the term have been various and conflicting. As Dr Helen Cam has remarked, the constitutional historian has tended to find the essence of Feudalism in the fact that 'landholding is the source of political power'; to the lawyer its essence has been that 'status is determined by tenure' and to the economic historian 'the cultivation of land by the exercise of rights over persons'.[1] But in general the matter has here excited little controversy. Definition has not been linked with rival social philosophies as has elsewhere been the case, most notably in nineteenth-century Russia. The very existence of such a system has not been called in question; and design for the future has not been made to depend on any imprint which this system may have left upon the present. In Russia, by contrast, the discussion has exercised opinion more powerfully than elsewhere, and the question whether Feudalism in the Western sense had ever existed formed a principal issue in the famous debate between Westerners and Slavophils in the first half and middle of

* From Maurice Dobb, *Studies in the Development of Capitalism* (Routledge & Kegan Paul, Ltd., 1946; International Publishers, 1947; 2nd edn., 1964), pp. 33–7. Reprinted by permission of Routledge & Kegan Paul and International Publishers Co. Inc., New York.

[1] *History*, vol. xxv (1940–1), p. 216.

the nineteenth century. At first emphasis was laid on the relationship in which the vassal stood to his prince or sovereign and on the form of landholding, yielding what was in the main a juridical definition: a definition certainly according with the etymology of the word, since as Maine observed the term Feudalism 'has the defect of calling attention to one set only of its characteristic incidents'. A matured example of this is the definition which the late Professor P. Struve recently contributed to the *Cambridge Economic History of Europe*: 'a contractual but indissoluble bond between service and land grant, between personal obligation and real right'. From this definition it followed that, although Feudalism had existed in Russia, its beginning was only to be dated from around 1350 with the termination of allodial landholding and the rise of service-tenures, and that it presumably terminated in the seventeenth century, when the *pomiestie* became assimilated to the *votchina* (i.e. became hereditary) and there was a reversion to the allodial principle.[2] With the growing influence of Marxism on Russian studies of agrarian history, a second type of definition came into prominence, giving pride of place to economic rather than to juridical relations. Professor M. N. Pokrovsky, for instance, who for many years was the *doyen* of Marxist historians, seems to have regarded Feudalism *inter alia* as a system of self-sufficient 'natural economy' by contrast with a moneyed 'exchange economy'—as 'an economy that has consumption as its object'.[3] This notion that Feudalism rested on natural economy as its economic base is one which, implicitly at least, seems to be shared by a number of economic historians in the West, and might be said to have more affinity with the conceptions of writers of the German Historical School, like Schmoller, than with those of Marx. There is a good deal of evidence to suggest that markets and money played a more prominent part in the Middle Ages than used to be

[2] *Cambridge Economic History of Europe*, vol. i, pp. 427, 432.
[3] *Brief History of Russia*, vol. i, p. 289. This definition *inter alia* earned him strong criticism from other Soviet historians in the early '30's. Pokrovsky's critics alleged that he tried simultaneously to ride both this conception and a purely political and juridical one; and that influenced in particular by a much-discussed work of Pavlov-Silvanski in 1907 (which championed the idea that Feudalism in the Western sense had existed in Russia), he never completely broke away from the latter conception (cf. S. Bakhrushin in *Protiv Historicheski Conseptsii M. N. Pokrovskovo*, pp. 117–18).

supposed. But this notion, at any rate, shares with the purely juridical one the great inconvenience (to say the least) of making the term not even approximately coterminous with the institution of serfdom. In Pokrovsky's case, for example, this definition leads him to speak of the sixteenth century in Russia as a period of decline of Feudalism (entitling the relevant chapter in his *Brief History* 'The Dissolution of Feudalism in Muscovy'), for the reason that commerce was reviving at this time and production for a market on the increase. Yet the sixteenth century was the very period when enserfment of previously free or semi-free peasants was taking place extensively and feudal burdens (in the common economic usage of the word) on the peasantry were being greatly augmented. Some English economic historians have apparently tried to evade this dilemma, first, by a virtual identification of serfdom with the performance of labour-services, or obligatory work directly performed upon the lord's estate, and, second, by attempting to show that such labour-services usually disappeared and were commuted into a contractual relationship in terms of money in the degree that trade and production for exchange in a wide market developed at the close of the Middle Ages. But this does not seem to provide at all a satisfactory way of escape, as what follows in this chapter will attempt to show.

The English mind is wont to dismiss arguments about definition as mere disputation about words: an instinct which is probably a healthy one seeing that so much argument of this kind has been little more than an exercise for pedants. But questions of definition cannot be entirely dismissed from our reckoning, however keen we may be on letting facts speak for themselves. We have already said that in attaching a definite meaning, whether explicitly or implicitly, to a term like Feudalism or Capitalism, one is *ipso facto* adopting a principle of classification to be applied in one's selection and assembly of historical events. One is deciding how one will break up the *continuum* of the historical process, the raw material that history presents to historiography—what events and what sequences are to be thrown into relief. Since classification must necessarily precede and form the groundwork for analysis, it follows that, as soon as one passes from description to analysis, the definitions one has adopted must have a crucial influence on the result.

To avoid undue prolixity, it must suffice, without further

parade of argument, to postulate the definition of Feudalism which in the sequel it is proposed to adopt. The emphasis of this definition will lie, not in the juridical relation between vassal and sovereign, nor in the relation between production and the destination of the product, but in the relation between the direct producer (whether he be artisan in some workshop or peasant cultivator on the land) and his immediate superior or overlord and in the social-economic content of the obligation which connects them. Conformably with the notion of Capitalism discussed in the previous chapter, this definition will characterize Feudalism primarily as a 'mode of production'; and this will form the essence of our definition. As such it will be virtually identical with what we generally mean by serfdom: an obligation laid on the producer by force and independently of his own volition to fulfil certain economic demands of an overlord, whether these demands take the form of services to be performed or of dues to be paid in money or in kind—of work or of what Dr Neilson has termed 'gifts to the lord's larder'.[4] This coercive force may be that of military strength, possessed by the feudal superior, or of custom backed by some kind of juridical procedure, or the force of law. This system of production contrasts, on the one hand, with slavery in that (as Marx has expressed it) 'the direct producer is here in possession of his means of production, of the material labour conditions required for the realization of his labour and the production of his means of subsistence. He carries on his agriculture and the rural house industries connected with it as an independent producer', whereas 'the slave works with conditions of labour belonging to another'. At the same time, serfdom implies that 'the property relation must assert itself as a direct relation between rulers and servants, so that the direct producer is not free': 'a lack of freedom which may be modified from serfdom with forced labour to the point of a mere tributary relation'.[5] It contrasts with Capitalism in that under the latter

---

[4] Neilson (1910), p. 15. Cf. Vinogradoff, p. 405: 'The labour-service relation, although very marked and prevalent in most cases [in the feudal period], is by no means the only one that should be taken into account.'

[5] *Capital*, vol. iii, p. 918. Marx goes on to say that 'under such conditions the surplus labour for the nominal owner of the land cannot be filched from them [the serfs] by any economic measures but must be forced from them by other measures, whatever may be the form assumed by them'; to which he adds the following remarks: 'The specific economic form in which unpaid surplus labour

the labourer, in the first place (as under slavery), is no longer an independent producer but is divorced from his means of production and from the possibility of providing his own subsistence, but in the second place (unlike slavery), his relation to the owner of the means of production who employs him is a purely contractual one (an act of sale or hire terminable at short notice): in the face of the law he is free both to choose his master and to change masters; and he is not under any obligation, other than that imposed by a contract of service, to contribute work or payment to a master. This system of social relations to which we refer as Feudal Serfdom has been associated in history, for a number of reasons, with a low level of technique, in which the instruments of production are simple and generally inexpensive, and the act of production is largely individual in character; the division of labour (and hence the co-ordination of individuals in production as a socially-integrated process) being at a very primitive level of development. Historically it has also been associated (and for a similar reason in the main) with conditions of production for the immediate needs of the household or village-community and not for a wider market; although 'natural economy' and serfdom are far from being coterminous, as we shall see. The summit of its development was characterized by demesne-farming: farming of the lord's estate, often on a considerable scale, by compulsory labourservices. But the feudal mode of production was not confined to this classic form. Finally, this economic system has been associated, for part of its life-history at least and often in its origins, with forms of political decentralization, with the conditional holding of land by lords on some kind of service-tenure, and (more generally) with the possession by a lord of judicial or quasi-judicial functions in relation to the dependent population. But, again, this association

---

is pumped out of the direct producers determines the relations of rulers and ruled. . . . It is always the direct relation of the owners of the conditions of production to the direct producers which reveals the innermost secret, the hidden foundation of the entire social construction, and . . . of the corresponding form of the state.' Yet 'this does not prevent the same economic basis from showing infinite variations and gradations in its appearance', due to 'numerous outside circumstances, natural environment, race peculiarities, outside historical influences, and so forth, all of which must be ascertained by careful analysis'.

is not invariable, and serfdom can be found in company both with fairly centralized State-forms and with hereditary landholding instead of service-tenures. To invert a description of Vinogradoff (who speaks of serfdom as 'a characteristic corollary of Feudalism'), we may say that the holding of land in fief is a common characteristic, but not an invariable characteristic, of Feudal Serfdom as an economic system in the sense in which we are using it.

FURTHER READING
*

Anderson, Perry *Passages from Antiquity to Feudalism* (1974)
Finley, M. I. *Ancient Slavery and Modern Ideology* (1980)
Hindess, Barry and Hirst, P. Q. *Pre-Capitalist Modes of Production* (1975)
Turner, Bryan S. *Marx and the End of Orientalism* (1978)

PART III

*

Social Classes

PRESENTATION

*

The concept of class has not only 'become the symbol of [Marx's] whole doctrine' (Ossowski, p. 99 below); it is one of the fundamental concepts of Marxism, in two respects. First, it was the historical starting point of Marx's own theory, in the sense that his discovery of the 'proletariat' led him, on one side, to seek the 'anatomy of civil society' in the mode of production, and, on the other side, to propound his distinctive view of the reconciliation of 'ideal' and 'real', in which the proletariat is conceived as being a real constitutive element of modern society, and at the same time an active force, the bearer of 'new values' and a 'new civilization'. Second, class is, in Marxist theory, the point at which the action of those structures—economic, political and ideological—which determine the character of a given mode of production and social formation, is concentrated and becomes effective (see Poulantzas 1973).

According to this conception every human society, from the dissolution of early communal society up to modern capitalism, is characterized by the existence of two principal classes—the masters of the conditions of production and the direct producers—and by a struggle between them over the control of the social labour process and the disposal of its product. In capitalist society, Marx and Engels argued in the *Communist Manifesto*, classes emerge with greater clarity, and the conflict between them becomes more acute, for whereas in earlier epochs 'we find almost everywhere a complicated arrangement of society into various orders, a manifold gradation of social rank', modern society is 'more and more splitting up into two great hostile camps, into two great classes directly facing each other—bourgeoisie and proletariat'. But capitalism is also destined to be the last 'antagonistic form of society', for just as the victory of the bourgeoisie abolished feudal estates, so the victory of the working class will abolish classes and inaugurate a classless society.

The Marxist theory of class has been, for almost a century now, a principal object of sociological criticism, revision, and defence. In support of its main thesis it can certainly be argued that during this period the political struggles in capitalist societies have been pre-eminently class struggles, and that the major political parties have been class parties. The criticisms of the thesis have taken three principal directions. First, it is claimed that the polarization of society between 'two great classes' has not taken place as Marx anticipated; on the contrary, the growth of the new middle classes has reintroduced that 'manifold gradation of social rank' characteristic of earlier epochs.[1] And partly for this reason, partly because of social reforms achieved within the capitalist system, which includes political democracy and the welfare state, the working class has failed to become a revolutionary force. The drive towards socialism, as the goal of a class movement, has lost much of its vigour.

Second, it has often been objected to the Marxist theory that it gives too much prominence to class struggles, while virtually ignoring other social divisions, forms of domination and subordination, and conflicts—such as those which arise from differences of gender, race, religion, or cultural distinctiveness—which have actually had a much greater practical significance, in some countries and during certain periods; and at the very least, have moderated or overshadowed the conflict between classes. In particular, it may be argued, as we have done in our Introduction (p. 10 above), that Marxist theory signally fails to take into account, or explain, the extraordinary vigour of nationalism, and the prevalence of conflict between nation states, which continue to have such a powerful impact upon the structure and development of human societies.

Finally, in the light of recent history, it seems open to question whether socialism would necessarily, or even probably, create a classless society, as the classical Marxist theory supposed; or whether new classes, new forms of domination and subordination, might not emerge—and have indeed already manifested themselves in the existing socialist societies (see Part VIII below)—on the basis of collective ownership of the means of production.

These themes will constitute a central preoccupation of Marxist sociology in the immediate future. The theory of class needs to be systematically reconstructed to take account of the

---

[1] See the text by Renner below, on the service class; Poulantzas (1975) on the new petty bourgeoisie; and Max Weber's analysis (1922) of the relation between class, status group, and party in the modern capitalist societies.

social and political action of other social groups, both in the economically advanced societies and in the countries of the Third World; to provide an analysis of the real development of the working class in the capitalist societies, in relation to other classes and groups, and to the nation state itself; and to establish a framework for explaining the formation of new group interests, and the growth of social conflict, in the present day socialist societies.

KARL MARX

*

*The Classes of Modern Society**

The owners of mere labour-power, the owners of capital, and the landowners, whose respective sources of income are wages, profit, and rent of land, or in other words, wage labourers, capitalists, and landowners, form the three great classes of modern society based on the capitalist mode of production.

The economic structure of modern society is indisputably most highly and classically developed in England. But even here the class structure does not appear in a pure form. Intermediate and transitional strata obscure the class boundaries even in this case, though very much less in the country than in the towns. However, this is immaterial for our analysis. We have seen that the constant tendency, the law of development of the capitalist mode of production, is to separate the means of production increasingly from labour, and to concentrate the scattered means of production more and more into large aggregates, thereby transforming labour into wage labour and the means of production into capital. There corresponds to this tendency, in a different sphere, the independent separation of landed property from capital and labour, or the transformation of all landed property into a form which corresponds with the capitalist mode of production.

The first question to be answered is—what constitutes a class? The answer can be found by answering another question: What constitutes wage labourers, capitalists, and landlords as the three great social classes?

At first glance it might seem that the identity of revenues and of sources of revenue is responsible. The classes are three great social groups whose components, the individual members, live from wages, profit and rent respectively, that is, from the utilization of their labour-power, capital and landed property.

However, from this point of view, doctors and officials would also form two distinct classes, for they belong to two different

* From Karl Marx, *Capital*, vol. iii, ch. 52. Translated by Tom Bottomore.

social groups, and the revenues of the members of each group come from the same source. The same would also be true of the infinite distinctions of interest and position which the social division of labour creates among workers as among capitalists and landowners; in the latter case, for instance, between owners of vineyards, farms, forests, mines, and fisheries. . . .

(Manuscript ends)

KARL MARX and FRIEDRICH ENGELS
*
*The Proletariat**

The proletariat goes through various stages of development. With its birth begins its struggle with the bourgeoisie. At first the contest is carried on by individual labourers, then by the workpeople of a factory, then by the operatives of one trade, in one locality, against the individual bourgeois who directly exploits them. They direct their attacks not against the bourgeois conditions of production, but against the instruments of production themselves; they destroy imported wares that compete with their labour, they smash to pieces machinery, they set factories ablaze, they seek to restore by force the vanished status of the workman of the Middle Ages.

At this stage the labourers still form an incoherent mass scattered over the whole country, and broken up by their mutual competition. If anywhere they unite to form more compact bodies, this is not yet the consequence of their own active union, but of the union of the bourgeoisie, which class, in order to attain its own political ends, is compelled to set the whole proletariat in motion, and is moreover yet, for a time, able to do so. At this stage, therefore, the proletarians do not fight their enemies, but the enemies of their enemies, the remnants of absolute monarchy, the landowners, the non-industrial bourgeois, the petty bourgeoisie. Thus the whole historical movement is concentrated in

---

* From Karl Marx and Friedrich Engels, *The Communist Manifesto* (1848). Text of the authorized English translation by Samuel Moore (1888).

the hands of the bourgeoisie; every victory so obtained is a victory for the bourgeoisie.

But with the development of industry the proletariat not only increases in number; it becomes concentrated in greater masses, its strength grows and it feels that strength more. The various interests and conditions of life within the ranks of the proletariat are more and more equalized, in proportion as machinery obliterates all distinctions of labour, and nearly everywhere reduces wages to the same low level. The growing competition among the bourgeois, and the resulting commercial crises, make the wages of the workers ever more fluctuating. The unceasing improvement of machinery, ever more rapidly developing, makes their livelihood more and more precarious; the collisions between individual workmen and individual bourgeois take more and more the character of collisions between two classes. Thereupon the workers begin to form combinations (trades unions) against the bourgeois; they club together in order to keep up the rate of wages; they found permanent associations in order to make provision beforehand for these occasional revolts. Here and there the contest breaks out into riots.

Now and then the workers are victorious, but only for a time. The real fruit of their battles lies, not in the immediate result, but in the ever-expanding union of the workers. This union is helped on by the improved means of communication that are created by modern industry, and that place the workers of different localities in contact with one another. It was just this contact that was needed to centralize the numerous local struggles, all of the same character, into one national struggle between classes. But every class struggle is a political struggle. And that union, to attain which the burghers of the Middle Ages, with their miserable highways, required centuries, the modern proletarians, thanks to railways, achieve in a few years.

This organization of the proletarians into a class, and consequently into a political party, is continually being upset again by the competition between the workers themselves. But it ever rises up again, stronger, firmer, mightier. It compels legislative recognition of particular interests of the workers, by taking advantage of the divisions among the bourgeoisie itself. Thus the Ten Hours Bill in England was carried.

Altogether, collisions between the classes of the old society further in many ways the course of development of the proletariat. The bourgeoisie finds itself involved in a constant battle. At first with the aristocracy; later on, with those portions of the bourgeoisie itself, whose interests have become antagonistic to the progress of industry; at all times with the bourgeoisie of foreign countries. In all these battles it sees itself compelled to appeal to the proletariat, to ask for its help, and thus, to drag it into the political arena. The bourgeoisie itself, therefore, supplies the proletariat with its own elements of political and general education, in other words, it furnishes the proletariat with weapons for fighting the bourgeoisie.

Further, as we have already seen, entire sections of the ruling classes are, by the advance of industry, precipitated into the proletariat, or are at least threatened in their conditions of existence. These also supply the proletariat with fresh elements of enlightenment and progress.

Finally, in times when the class struggle nears the decisive hour, the process of dissolution going on within the ruling class, in fact within the whole range of old society, assumes such a violent, glaring character, that a small section of the ruling class cuts itself adrift, and joins the revolutionary class, the class that holds the future in its hands. Just as, therefore, at an earlier period, a section of the nobility went over to the bourgeoisie, so now a portion of the bourgeoisie goes over to the proletariat, and in particular, a portion of the bourgeois ideologists, who have raised themselves to the level of comprehending theoretically the historical movement as a whole.

KARL MARX
*

*The Petty Bourgeoisie and the Peasantry in France**

Against the coalition of the bourgeoisie, a coalition between petty bourgeois and workers had been formed, the so-called *social-democratic* party. The petty bourgeois saw themselves as being ill-

---

* From Karl Marx, *The Eighteenth Brumaire of Louis Bonaparte* (1852), sects iii, vii. Translated by Tom Bottomore.

rewarded after the June days of 1848,[1] their material interests imperilled, and the democratic guarantees which were to ensure the implementation of these interests called in question by the counter-revolution. Hence they drew closer to the workers. (. . .) The peculiar character of Social-Democracy is epitomized in the fact that democratic-republican institutions are demanded as a means, not of abolishing the two extremes, capital and wage labour, but of attenuating their antagonism and transforming it into harmony. However varied the means that may be proposed for accomplishing this end, however much it may be embellished with more or less revolutionary ideas, the content remains the same. This content is the transformation of society in a democratic way, but a transformation within the limits of the petty bourgeoisie. Only one must not form the narrow-minded notion that the petty bourgeoisie wants to promote, on principle, an egoistic class interest. Rather, it believes that the *particular* conditions of its emancipation are the *general* conditions within which alone modern society can be saved and the class struggle avoided. Just as little must one suppose that the democratic representatives are indeed all shopkeepers or enthusiastic champions of shopkeepers. According to their education and their individual position they may be as far apart as heaven and earth. What makes them representatives of the petty bourgeoisie is the fact that in their minds they do not go beyond the limits which the latter do not overstep in life; that they are consequently driven, theoretically, to the same problems and solutions to which material interest and social position drive the latter in practice. This is, in general, the relation of the *political* and *literary representatives* of a class to the class they represent. (. . .)

And yet the state power does not float in mid-air. Bonaparte represents a class, and indeed the most numerous class of French society, the smallholding peasants. Just as the Bourbons were the dynasty of large landed property, and the Orleans the dynasty of money, so the Bonapartes are the dynasty of the peasants, that is, of the mass of the French people. (. . .)

The smallholding peasants from a vast mass, the members of which live in similar conditions but without entering into

[1] 23–6 June 1848, when General Cavaignac, the Minister of War, savagely repressed the revolt of the Parisian workers, at a cost of at least 1,500 dead and 12,000 arrested, many of whom were subsequently deported to Algeria [Eds.].

manifold relations with each other. Their mode of production isolates them from one another instead of bringing them into mutual intercourse. The isolation is furthered by France's poor means of communication and by the poverty of the peasants. Their sphere of production, the smallholding, permits no division of labour in its cultivation, no application of science, and hence no diversity of development, no variety of talent, no wealth of social relationships. Each individual peasant family is almost self-sufficient, produces directly the major part of its own consumption, and thus acquires its means of life more through an exchange with nature than in intercourse with society. The smallholding, the peasant and his family; alongside them another smallholding, another peasant and another family. Three score of these make up a village, and three score villages make up a Department. In this way, the great mass of the French nation is formed by the simple addition of homologous magnitudes, much as potatoes in a sack form a sack of potatoes. In so far as millions of families live under economic conditions of existence that separate their mode of life, their interests, and their culture from those of the other classes, and put them in hostile opposition to the latter, they form a class. In so far as there is merely a local interconnection among these smallholding peasants, and the identity of their interests begets no community, no national bond, and no political organization among them, they do not form a class. They are consequently incapable of implementing their class interests in their own name, whether through a parliament or through a convention. They cannot represent themselves; they must be represented. Their representative must appear at the same time as their master, as an authority over them, as an unlimited governmental power which protects them against the other classes and sends them rain and sunshine from above. The political influence of the smallholding peasants, therefore, finds its ultimate expression in the executive power subordinating society to itself. (. . .)

The economic development of smallholding property has changed fundamentally the relation of the peasants to the other classes of society. Under Napoleon, the fragmentation of land and soil in the countryside complemented free competition and the emerging large scale industry in the towns. The peasant class was the ubiquitous protest against the landed aristocracy which had just been overthrown. The roots which smallholding property struck in French soil deprived feudalism of all nutriment. Its boundary posts constituted the natural fortifications of the bourgeoisie against any surprise attack by its old overlords. But in the course of the nineteenth century the urban usurer took the place of the feudal one, the mortgage replaced the feudal obligation that went with the land, and aristocratic landed property was replaced by bourgeois capital. The smallholding of the peasant is now only the pretext which allows the capitalist to draw profits, interest, and rent from the soil, while leaving it to the peasant cultivator himself to see how he can extract his wages. The mortgage debt burdening the soil of France imposes on the French peasantry an amount of interest equal to the annual interest on the entire British national debt. Smallholding property, in this enslavement by capital to which its development inevitably leads, has transformed the mass of the French nation into troglodytes. Sixteen million peasants (including women and children) live in hovels, a large number of which have only one opening, others only two, and the most favoured only three. But windows are to a house what the five senses are to the head. The bourgeois order, which at the beginning of the century set the state as a sentinel to guard the newly arisen smallholding and manured it with laurels, has become a vampire which sucks out its life blood and brain marrow and throws them into the alchemist's cauldron of capital. The *Code Napoléon* is now only a *codex* of distraints, forced sales, and compulsory auctions. To the four million (including children, etc.) officially recognized paupers, vagabonds, criminals, and prostitutes in France must be added another five million who hover on the margin of existence and either drag out their lives miserably in the countryside itself, or, with their rubbish and their children, continually leave the countryside for the towns and the towns for the countryside. The interests of the peasants, therefore, are no longer, as under Napoleon, in accord with, but in opposition to, the interests of the bourgeoisie, to capital. Hence they find their natural ally and leader in the *urban proletariat*, whose task is the overthrow of the bourgeois order.

STANISLAW OSSOWSKI

*

Marx's Concept of Class*

The concept of social class is something more than one of the fundamental concepts of Marxian doctrine. It has in a certain sense become the symbol of his whole doctrine and of the political programme that is derived from it. This concept is expressed in the terms 'class standpoint' and 'class point of view,' which in Marxist circles used until recently to be synonymous with 'Marxist standpoint' or 'Marxist point of view'. In this sense 'class standpoint' simply meant the opposite of 'bourgeois standpoint'.

According to Engels,[1] Marx effected a revolutionary change in the whole conception of world history. For Marx, so Engels maintained, had proved that 'the whole of previous history is a history of class struggles, that in all the simple and complicated political struggles the only thing at issue has been the social and political rule of social classes'.

The concept of social class is also linked with what Engels in the same article calls the second great discovery of Marx, to which he attaches so much importance in the history of science—the clarification of the relationship that prevails between capital and labour. Finally, it may be said that the concept of social class is bound up with the entire Marxian conception of culture as the superstructure of class interests.

The role of the class concept in Marxian doctrine is so immense that it is astonishing not to find a definition of this concept, which they use so constantly, anywhere in the works of either Marx or Engels. One might regard it as an undefined concept of which the meaning is explained contextually. But in fact one has only to compare the various passages in which the concept of social class

---

* From Stanislaw Ossowski, *Class Structure in the Social Consciousness* (New York: The Free Press, 1963; London: Routledge & Kegan Paul Ltd., 1963), pp. 70–4. Copyright © 1963 by Stanislaw Ossowski. Reprinted with the permission of Routledge & Kegan Paul Ltd.

---

[1] ME, vol. ii, p. 149; the quotation comes from F. Engels, *Karl Marx*. [ME refers to Karl Marx and Frederick Engels, *Selected Works in Two Volumes* (Moscow: Foreign Languages Publishing House, 1951) Eds.].

is used by either writer to realize that the term 'class' has for them a variable denotation: that is, that it refers to groups differentiated in various ways within a more inclusive category, such as the category of social groups with common economic interests, or the category of groups whose members share economic conditions that are identical in a certain respect. The sharing of permanent economic interests is a particularly important characteristic of social classes in Marxian doctrine, and for this reason it has been easy to overlook the fact that although it is, in the Marxian view, a *necessary condition* it does not constitute a *sufficient condition* for a valid definition of social class.

Marx left the problem of producing a definition of the concept of social class until much later. The manuscript of the third volume of his *magnum opus, Das Kapital*, breaks off dramatically at the moment when Marx was about to answer the question: 'What constitutes a class?' We do not know what answer he would have given if death had not interrupted his work. Nor do we know whether he would have attempted to explain the discrepancies in his earlier statements.

After the death of Marx, Engels did not take up the question which the manuscript of *Das Kapital* left unanswered. Lenin's later definition, which has been popularized by Marxist textbooks and encyclopaedias, links two different formulations but fails to explain how we are to regard them. Does the author see them as two equivalent definitions and does he link them in order to give a fuller characteristic of the designate of the concept of class? Or is the conjunction of the two formulations essential because the characteristics given in one of them are not necessarily conjoint to the characteristic given in the second? Independently of this, such metaphorical expressions as the 'place in the historically determined system of social production' may be variously interpreted and Lenin's definition is sufficiently loose to be applicable to all the shades of meaning found in the term 'class' as used by Marx and Engels.[2] Bukharin's definition,

---

[2] 'Classes are large groups of people which differ from each other by the place they occupy in a historically determined system of social production, by their relation (in most cases fixed and formulated in law) to the means of production, by their role in the social organization of labour and, consequently, by the dimensions and method of acquiring the share of social wealth of which they dispose. Classes are groups of people one of which can appropriate the labour of another owing to the different places they occupy in a definite system of social

which is also intended to reflect the Marxian conception of social class,[3] affords room for even wider possibilities of interpretation, and it is only Bukharin's classification of social classes that enables one to grasp the denotation assigned by the author to the concept of social class.[4]

In using the concept of class based on economic criteria, Marx sometimes restricts the scope of this concept by introducing psychological criteria. An aggregate of people which satisfies the economic criteria of a social class becomes a class in the full meaning of this term only when its members are linked by the tie of class consciousness, by the consciousness of common interests, and by the psychological bond that arises out of common class antagonisms.[5] Marx is aware of the ambiguity and makes a terminological distinction between *Klasse an sich* and *Klasse für sich*, but he does not in general make much further use of these more narrowly defined concepts.

Marx sometimes uses a different term to denote a class which is not a class in the fullest sense because it lacks psychological bonds. For instance, he sometimes uses the term 'stratum'; on other occasions he avoids using a more general term and confines himself to the name of a specified group such as the 'small

economy.' (V. I. Lenin, *A Great Beginning*, in *The Essentials of Lenin* in Two Volumes, London, Lawrence & Wishart, 1947, p. 492.)

[3] N. Bukharin, *Historical Materialism, A System of Sociology*, London, 1926, p. 267 (English translation).

[4] Ibid., pp. 282-4.

[5] Cf. the following passages:

'The separate individuals form a class in so far as they have to carry on a common battle against another class.' (K. Marx and F. Engels, *The German Ideology* (The Marxist-Leninist Library, vol. xvii, London, Lawrence & Wishart, 1940, pp. 48-9).) 'The organization of the proletarians into a class, and consequently into a political party.' ('Manifesto of the Communist Party,' ME, Vol. i, p. 41.) 'In so far as millions of families live under economic conditions of existence that separate their mode of life, their interests and their culture from those of the other classes, and put them in hostile opposition to the latter, they form a class. In so far as there is merely a local interconnection among these small-holding peasants, and the identity of their interests begets no community, no national bond and no political organization among them, they do not form a class. They are consequently incapable of enforcing their class interest.' (ME, vol. i, p. 303; quotation from K. Marx.) 'Bonaparte represented the most numerous class of the French society at that time, the small-holding (*Parzellen*) peasants' (ME, vol. i, p. 302; quotation from 'The Eighteenth Brumaire of Louis Bonaparte').

peasantry'. At times he may even call certain classes which are conscious of their class interests 'fractions' of a more inclusive class. In the case of capitalists and landowners, for instance, Marx sometimes sees them as two separate classes, at others as two fractions of a single class, the bourgeoisie.

All these discrepant uses of the term 'class' were probably the less important for Marx because, according to his theory, further social development would render them obsolete. This was to result from the growth of the social consciousness and from the predicted disappearance of the difference between the *Klasse an sich* and the *Klasse für sich* as well as from the progressive process of class polarization in the social structure.

The matter can however be put in a different way. We may take it that Marx, instead of providing a definition of social class which would make it possible to fix the scope of this concept, is giving the model of a social class, the ideal type which is to be fully realized in the future, in the last stage of the development of the capitalist system. In the period in which Marx wrote, the industrial proletariat of Western Europe was approximating to the ideal type of a social class. Other social groups separated on the basis of economic criteria could be called classes only to a greater or lesser extent, and could approximate to the ideal type only in some respects. Hence endeavours to apprehend them by means of conceptual categories with sharply-drawn boundaries of application must lead to confusion.

However that may be, one should, when considering the Marxian conception of class structure, remember that the component elements of this structure are confined to those groups which Marx calls 'classes' when contrasting them with 'strata', in which 'the identity of their interests (those of the members of a "stratum") begets no unity, no national union and no political organization.'

KARL KAUTSKY

*

Class, Occupation, and Status*

The concept of status group is related to the concepts of class and occupation, but is not identical with either of them. I understand by it a group of members of a community who are distinguished from the other members by explicit definitions formulated by the community. This happens as a result of particular rights being assigned to, or duties imposed upon, the group; and membership of the group is linked to specific conditions, such as inheritance or proof of specific kinds of knowledge, and so on.

In order not to get lost in too many details, let us disregard caste, which like a status group indicates a group of people who are especially privileged or oppressed, and is a particularly 'closed' group, though this is the result of old fossilized customs which have assumed a sacred and inviolable character, rather than of any deliberate regulation by the community. Originally the castes may have been merely status groups, which emerged from community decisions. But this origin has been forgotten; the community life which created them has itself perished and has been replaced by new structures, but the status division has remained. A status group which was created or given recognition by specific community laws can be abolished by new laws, but castes can only be abolished by overcoming the power of old customs. Just like the English, their conquerors and successors from the 17th century, the Muhammadan tribes which invaded India, and from the 11th century established new states on the ruins of the old, regarded the caste system which they encountered as an absurdity. Yet it still continues to exist. Caste is a particular kind of status. Status itself, at least in its origins, always represents a particular occupation or a particular class, or a combination of specific occupations or classes. Thus the Third Estate in France, up to the Revolution, embraced capitalists and wage workers, artisans and peasants, as well as intellectuals; that is, the great mass of the population. The separation of particular sections of a people from others by status can to some extent be

---

* From Karl Kautsky, *Die materialistische Geschichtsauffassung* (Berlin: J. H. W. Dietz Nachf., 1927), vol. 2, pp. 38–42. Translated by Patrick Goode. Published by permission of Verlag J. H. W. Dietz Nachf.

imposed on these sections by the community or its ruling class, in order to maintain an existing order of society. But it can also be imposed on the community by the elements which have grouped and separated themselves by status, in order to consolidate their victory over other sections of the community.

Furthermore, different status groups have very different characteristics according to whether they correspond to an occupation or a class. The status group of doctors, or that of government officials, is quite different from that of the nobility. At all events, when an occupation is given the character of a status group this usually means that it is also provided with specific privileges and monopolies, which make its position approximate to that of a class. Thus in France (and in many other countries before the Revolution) we find a nobility of office (*noblesse de robe*) emerging in addition to a nobility of birth (*noblesse d'épée*). In many countries down to our own times the privileged position of the officer stratum often gives it the character of a ruling and exploiting class. This is also true of the clergy of many religions. In this status group the boundary line between occupation and class is very often unclear.

At its origins a status group coincides with a definite class (or set of classes) or a definite occupation, but this rarely remains the case. We have already seen that in human society we have to distinguish between fixed and fluid elements. The latter are the driving forces of the development of society; the former are obstacles to it. Every regulation of a social relationship or social phenomenon, once it has been established and generally recognized, has the tendency to become conservative and rigid. But social life, which created these relationships and phenomena, continues and creates conditions which contradict the arrangements which were fixed and generally recognized. Thus speech is not rigid. Not only the meaning of invidivual words, but also their pronunciation, changes in the course of centuries. Writing is much more conservative than speech. The recognized way of writing words tends to remain unchanged, and this can lead to enormous divergences between the written and the spoken word, as the English language shows most strikingly.

We have also seen that the family, which emerges directly from the production process, is constantly being changed as a result of changes in the latter, while the established kinship systems are, as Morgan puts it, 'passive': 'Only over long periods do they register the progress which the family has made in the course of time, and they only undergo a radical change when the family itself has radically changed.'

Property relations are also a passive element, as compared with technology and the production process. The same holds true of the relationship between status group and occupation or class.

Technology and production change uninterruptedly, and along with them occupations and classes. But the privileges, burdens, and circumstances of the members of individual status groups do not change to the same extent. On one side this may lead to the institutions of a status group becoming increasingly restrictive for the process of production and increasingly oppressive for society; and on the other side, to a growing attrition of the original identity between a status group and the corresponding class or occupation. Thus, for example, the feudal nobility was originally synonymous with the class of large landed proprietors. But what eventually happened in the course of time, even before the collapse of feudalism, was that non-noble elements acquired extensive landed property, while on the other hand, not only individual nobles, but entire noble families, lost their landed property. Such families were obliged either to look for another way of exploiting the mass of the people (for example, as well-paid idle courtiers) or to sink into the mass of the exploited themselves.

On the other side, from the time when the migration of peoples ended until well into the Middle Ages, the status group of the Catholic clergy was to a great extent identical with the profession of intellectual. But as urban life, and commerce with the Orient after the Crusades, developed, so the intellectuals who did not belong to the clergy became increasingly numerous, until finally the functions of intellectuals in the modern sense came to be carried on almost entirely outside the Catholic clergy, with a few exceptions (. . .).

Thus a status group does not by any means always coincide with the class or the occupation of which it was originally a petrified form. Yet its social role always has much in common with that of a class or an occupation. Hence class struggles very often assume the form of conflicts between status groups. The *Communist Manifesto*, which begins with the statement that all recorded history is the history of class struggles, illustrates this proposition in the following manner: 'Freeman and slave, patrician and plebeian, lord and serf, guild-master and journeyman, in a word, oppressor and oppressed, stood in constant opposition to one another, carried on an uninterrupted, now hidden, now open fight, a fight that each time ended, either in a revolutionary reconstitution of society at large, or in the common ruin of the contending classes.' I shall leave aside the question as to how far the conflicts between guild-masters and journeymen, in the early period and the maturity of handicrafts, before the emerging industrial capitalism began to react upon handicraft production, should be regarded as class conflicts. Both the journeyman and the apprentice were on the way to becoming masters. The future master was already prefigured in the journeyman, and they represented simply a different stage in the life of the same class. The journeyman belonged to the same class as the master, just as the caterpillar belongs to the same species as the butterfly. With the decline of the guilds there emerges a class opposition between journeymen and masters, for not every journeyman now becomes a master. In so far as an opposition between masters and journeymen arose, it was one between two status groups. The powers and duties of both masters and journeymen were strictly defined by law.

The other classes referred to in this passage of the *Communist Manifesto* were also status groups, and its authors were quite aware of this fact. Immediately after the passage just cited they say: 'In the earlier epochs of history, we find almost everywhere a complicated arrangement of society into various orders, a manifold gradation of social rank. In ancient Rome we have patricians, knights, plebeians, slaves; in the Middle Ages, feudal lords, vassals, guild-masters, journeymen, apprentices, serfs; in almost all of these classes, again, subordinate gradations.' Here Marx and Engels spoke of the same groups sometimes as status groups, sometimes as classes. In spite of this, they were highly critical of Lassalle for referring to wage earners, without any distinction, sometimes as a class and sometimes as a status group, the Fourth Estate; and they were entirely justified in their criticism. Every status group, in so far as it does not represent a fossilized occupation, is a fossilized class, and a status group of the latter kind can also be termed a class. But it is not the case, conversely, that every class is a status group. The latter is only a particular phenomenal form of some classes.

In pre-capitalist society, which developed extremely slowly in every respect, classes could take on the rigidity of a status group and maintain it over a long period. Capitalist society, which undergoes uninterrupted technical, economic, political, and scientific transformation, is incompatible with the rigidity of any kind of organization by status groups. It dissolves all status bonds, as it has already begun to undermine the very deep-rooted and tenacious caste distinctions in India. Capitalist society only tolerates classes without the disguise and constriction of status. This is the distinguishing feature of capitalism, just as that of socialism will be the abolition of all classes. Hence Lassalle was quite mistaken in speaking of the proletariat as the Fourth Estate.

Where status groups or professional associations exist, they are quite apparent. The same is not true of classes, which are not sharply defined by the community. Hence, when the bourgeois revolutions abolished the Estates system their champions and ideologues thought that the reign of liberty, equality, and fraternity had begun. Individual differences between rich and poor, educated and ignorant, were still apparent, but people were convinced that good schools and progressive taxes on income, property and inheritance were bound to remedy these differences and finally cause them to disappear.

GEORG LUKÁCS

*

*Class Consciousness**

Bourgeoisie and proletariat are the only pure classes in bourgeois society. They are the only classes whose existence and development are entirely dependent on the course taken by the modern evolution of production and only from the vantage point of these classes can a plan for the total organization of society *even be imagined*. The outlook of the other classes (petty bourgeois or peasants) is ambiguous or sterile because their existence is not based exclusively on their role in the capitalist system of production but is indissolubly linked with the vestiges of feudal

* From Georg Lukács, *History and Class Consciousness*. Translated by Rodney Livingstone (London: Merlin Press, 1971), pp. 59–62, 68–9, 73, 80. Reprinted by permission of The Merlin Press Ltd.

society. Their aim, therefore, is not to advance capitalism or to transcend it, but to reverse its action or at least to prevent it from developing fully. Their class interest concentrates on *symptoms of development* and not on development itself, and on elements of society rather than on the construction of society as a whole.

The question of consciousness may make its appearance in terms of the objectives chosen or in terms of action, as for instance in the case of the petty bourgeoisie. This class lives at least in part in the capitalist big city and every aspect of its existence is directly exposed to the influence of capitalism. Hence it cannot possibly remain wholly unaffected by the *fact* of class conflict between bourgeoisie and proletariat. But as a 'transitional class in which the interests of two other classes become simultaneously blunted . . .' it will imagine itself 'to be above all class antagonisms'.[1] Accordingly it will search for ways whereby it will 'not indeed eliminate the two extremes of capital and wage labour, but will weaken their antagonism and transform it into harmony'.[2] In all decisions crucial for society its actions will be irrelevant and it will be forced to fight for both sides in turn but always without consciousness. In so doing its own objectives—which exist exclusively in its own consciousness—must become progressively weakened and increasingly divorced from social action. Ultimately they will assume purely 'ideological' forms. The petty bourgeoisie will only be able to play an active role in history as long as these objectives happen to coincide with the real economic interests of capitalism. (. . .)

This isolation from society as a whole has its repercussions on the internal structure of the class and its organizational potential. This can be seen most clearly in the development of the peasantry. Marx says on this point:[3]

The small-holding peasants form a vast mass whose members live in similar conditions but without entering into manifold relations with each other. Their mode of production isolates them from one another instead of bringing them into mutual intercourse. . . . Every single peasant family . . . thus acquires its means of life more through exchange with nature than in intercourse with society. . . . In so far as millions of families live under economic conditions of existence that

[1] Marx, *The Eighteenth Brumaire of Louis Bonaparte*.
[2] Ibid.
[3] Ibid.

separate their mode of life, their interests and their culture from those of other classes and place them in opposition to them, they constitute a class. In so far as there is only a local connection between the smallholding peasants, and the identity of their interests begets no community, no national unity and no political organization, they do not constitute a class.

Hence *external* upheavals, such as war, revolution in the towns, etc. are needed before these masses can coalesce in a unified movement, and even then they are incapable of organizing it and supplying it with slogans and a positive direction corresponding to their own interests.

Whether these movements will be progressive (as in the French Revolution of 1789 or the Russian Revolution of 1917), or reactionary (as with Napoleon's *coup d'état*) will depend on the position of the other classes involved in the conflict, and on the level of consciousness of the parties that lead them. For this reason, too, the *ideological* form taken by the class-consciousness of the peasants changes its content more frequently than that of other classes: this is because it is always borrowed from elsewhere. Hence parties that base themselves wholly or in part on this class consciousness always lack really firm and secure support in critical situations (as was true of the Socialist Revolutionaries in 1917 and 1918). This explains why it is possible for peasant conflicts to be fought out under opposing flags. (. . .)

The distinction between the two modes of contradiction may be briefly described in this way: in the case of the other classes, a class-consciousness is prevented from emerging by their position within the process of production and the interests this generates. In the case of the bourgeoisie, however, these factors combine to produce a class-consciousness but one which is cursed by its very nature with the tragic fate of developing an insoluble contradiction at the very zenith of its powers. As a result of this contradiction it must annihilate itself.

The tragedy of the bourgeoisie is reflected historically in the fact that even before it had defeated its predecessor, feudalism, its new enemy, the proletariat, had appeared on the scene. Politically, it became evident when at the moment of victory, the 'freedom' in whose name the bourgeoisie had joined battle with feudalism, was transformed into a new repressiveness. Sociologically, the bourgeoisie did everything in its power to eradicate the fact of class conflict from the consciousness of society, even though class conflict had only emerged in its purity and became established as an historical fact with the advent of capitalism. Ideologically, we see the same contradiction in the fact that the bourgeoisie endowed the individual with an unprecedented importance, but at the same time that same individuality was annihilated by the economic conditions to which it was subjected, by the reification created by commodity production.

All these contradictions, and the list might be extended indefinitely, are only the reflection of the deepest contradictions in capitalism itself as they appear in the consciousness of the bourgeoisie in accordance with their position in the total system of production. For this reason they appear as dialectical contradictions in the class consciousness of the bourgeoisie. They do not merely reflect the inability of the bourgeoisie to grasp the contradictions inherent in its own social order. For, on the one hand, capitalism is the first system of production able to achieve a total economic penetration of society,[4] and this implies that in theory the bourgeoisie should be able to progress from this central point to the possession of an (imputed) class-consciousness of the whole system of production. On the other hand, the position held by the capitalist class and the interests which determine its actions ensure that it will be unable to control its own system of production even in theory. (. . .)

The unique function of consciousness in the class struggle of the proletariat has consistently been overlooked by the vulgar Marxists who have substituted a petty 'Realpolitik' for the great battle of principle which reaches back to the ultimate problems of the objective economic process. Naturally we do not wish to deny that the proletariat must proceed from the facts of a given situation. But it is to be distinguished from other classes by the fact that it goes beyond the contingencies of history; far from being driven forward by them, it is itself their driving force and impinges centrally upon the process of social change. When the vulgar Marxists detach themselves from this central point of

[4] But no more than the tendency. It is Rosa Luxemburg's great achievement to have shown that this is not just a passing phase but that capitalism can only survive—economically—while it moves society in the direction of capitalism but has not yet fully penetrated it. This economic self-contradiction of any purely capitalist society is undoubtedly one of the reasons for the contradictions in the class consciousness of the bourgeoisie.

view, i.e. from the point where a proletarian class-consciousness arises, *they thereby place themselves on the level of consciousness of the bourgeoisie*. And that the bourgeoisie fighting on its own ground will prove superior to the proletariat both economically and ideologically can come as a surprise only to a vulgar Marxist. Moreover only a vulgar Marxist would infer from this fact, which after all derives exclusively from his own attitude, that the bourgeoisie *generally* occupies the stronger position. For quite apart from the very real force at its disposal, it is self-evident that the bourgeoisie *fighting on its own ground* will be both more experienced and more expert. Nor will it come as a surprise if the bourgeoisie automatically obtains the upper hand when its opponents abandon their own position for that of the bourgeoisie.

As the bourgeoisie has the intellectual, organizational and every other advantage, the superiority of the proletariat must lie exclusively in its ability to see society from the centre, as a coherent whole. This means that it is able to act in such a way as to change reality; in the class-consciousness of the proletariat theory and practice coincide and so it can consciously throw the weight of its actions on to the scales of history—and this is the deciding factor. When the vulgar Marxists destroy this unity they cut the nerve that binds proletarian theory to proletarian action. They reduce theory to the 'scientific' treatment of the symptoms of social change and as for practice they are themselves reduced to being buffeted about aimlessly and uncontrollably by the various elements of the process they had hoped to master. (. . .)

The dialectical cleavage in the consciousness of the proletariat is a product of the same structure that makes the historical mission of the proletariat possible by pointing forward and beyond the existing social order. In the case of the other classes we found an antagonism between the class's self-interest and that of society, between individual deed and social consequences. This antagonism set an external limit to consciousness. Here, in the centre of proletarian class-consciousness we discover an antagonism between momentary interest and ultimate goal. The outward victory of the proletariat can only be achieved if this antagonism is inwardly overcome.

As we stressed in the motto to this essay the existence of this conflict enables us to perceive that class-consciousness is identical with neither the psychological consciousness of individual members of the proletariat, nor with the (mass-psychological) consciousness of the proletariat as a whole; but it is, on the contrary, *the sense, become conscious, of the historical role of the class*. This sense will objectify in particular interests of the moment which may only be omitted at the price of allowing the proletarian class struggle to slip back into the most primitive Utopianism. Every momentary interest may have either of two functions: either it will be a step towards the ultimate goal or else it will conceal it. Which of the two it will be depends *entirely upon the class consciousness of the proletariat and not on victory or defeat in isolated skirmishes.* (. . .)

In view of the great distance that the proletariat has to travel ideologically it would be disastrous to foster any illusions. But it would be no less disastrous to overlook the forces at work within the proletariat which are tending towards the ideological defeat of capitalism. Every proletarian revolution has created workers' councils in an increasingly radical and conscious manner. When this weapon increases in power to the point where it becomes the organ of state, this is a sign that the class-consciousness of the proletariat is on the verge of overcoming the bourgeois outlook of its leaders.

The revolutionary workers' council (not to be confused with its opportunist caricatures) is one of the forms which the consciousness of the proletariat has striven to create ever since its inception. The fact that it exists and is constantly developing shows that the proletariat already stands on the threshold of its own consciousness and hence on the threshold of victory. The workers' council spells the political and economic defeat of reification. In the period following the dictatorship it will eliminate the bourgeois separation of the legislature, administration and judiciary. During the struggle for control its mission is twofold. On the one hand, it must overcome the fragmentation of the proletariat in time and space, and on the other, it has to bring economics and politics together into the true synthesis of proletarian praxis. In this way it will help to reconcile the dialectical conflict between immediate interests and ultimate goal.

KARL RENNER

*

The Service Class*

In addition to those changes already mentioned there is another which is no less important. In economic terms, the capitalist, *qua* capitalist, is an agent of circulation, and as such he makes use of paid assistants who, as we have explained, gradually take over his functions. These agents are not wage workers; they do not produce, but dispose of the values that have been produced. As long as a capitalist economy exists their services are also socially necessary.[1] The capitalist performs these services for profit, the manager for a salary, a fee, or a share in the profits (commission).

1. *The Public Service*. The public service has served capital as a model for this arrangement. Historically the state has had a dual function: external and internal security (the military and police); and the creation and administration of law (the legislature and judiciary). In the feudal period, with its undeveloped money economy, the state officials were paid by investing them with the ownership of land and serfs, and priests were paid in a similar way by the church. It was the town burgesses and the bourgeois state which first began to pay salaries to their agents.

A salary differs fundamentally by its nature and in the way it is assessed from a wage. What the salaried person produces is not a commodity, not an 'economic' good, but law and order. His work is most effective when law and order are made so secure that he has very little to do. The best gendarme is the one who is effective merely by existing, like the judge whose established authority limits the number of disputed cases. The payment of a salary is not meant to provide food and shelter from week to week but to establish a life style and family situation which improves as the individual becomes older and terminates with an old age

---

* From *Wandlungen der modernen Gesellschaft* (1953), pp. 211–14. English translation from *Austro-Marxism*, texts translated and edited by Tom Bottomore and Patrick Goode (Oxford: OUP, 1978). Copyright © Tom Bottomore and Patrick Goode, 1978. Reprinted by permission of Oxford University Press.

---

[1] In a future social order they will become agents of distribution instead of agents of circulation, and as we have shown, distribution follows different laws from those which regulate circulation.

pension. The salary is used for consumption and procures commodities for the consumer without any economic counterpart, that is to say in exchange for money which is not acquired in the economic process, but is forcibly extracted from the latter by government taxes and duties. Thus salaries represent a further diversion of the surplus value of society by means of direct or indirect taxation.

The active capitalist pays his helpers in the same manner, and in so far as he himself ceases to be active he pays substitutes as well as assistants, all those agents of management to whom he refers quite properly not as workers but as officials or employees.

2. *The Private Employee.* It is not only the state and economic enterprises, but also all the new associations developed alongside the real economic process (especially in the sphere of circulation), that create their executive agents in the same manner. For the most part they establish this 'employee relationship' by means of a contract which does not create a relationship of wage labour, although at first it is partly constructed on analogous lines (weekly or monthly payment, agreements concerning length of service, conditions of dismissal). The more explicitly the service relationship becomes one of trust, the more it tends to become firmly established. The code of service as a more or less hierarchical norm replaces the labour contract. The position is officially confirmed and dismissal is only possible in specific circumstances, for the most part only by the decision of an arbitrator or judge.

A brief survey of the preceding development reveals the remarkable extension of this kind of legal relationship and the extraordinary numerical increase in the class of people whose existence is regulated by it. The service class has emerged alongside the working class in the strictly technical sense. The expression 'service class' marks a fundamental distinction from the traditional 'serving class', which performed real labour, if only at a rudimentary technical level, and was for the most part paid in kind. Three basic forms can be distinguished: economic service (managers, etc.), social service (distributive agents of welfare services), and public service (public, official agents).

3. *From a Caste to a Class.* Traditionally these new social strata are at first both materially and intellectually opposed to the working class. Since for the most part they are recruited from the bourgeois middle class, and since their salary supplements their

own income from property, they could not be classified with the propertyless class. Initially they formed what might be called in sociological terms a caste (the priestly caste, the intellectuals, the officer caste, the caste of officials). These castes are separated from the general social milieu by particular qualifications, and set in opposition to society. It is a characteristic feature of these castes that they extend through all ranks and strata of society; from the Generalissimo down to the common soldier, from the cardinal to the village chaplain, from the head of the office down to the office boy, and so on. The same caste spirit imbues both high and low with the love of rank and of exclusiveness and sets them in opposition to the people as a whole. From this point of view, historical revolutions are very often the consequence of a struggle of the whole people against a ruling caste or of a struggle for supremacy between different castes (the secular bureaucracy against the ecclesiastical hierarchy, civil against military).

Before the Second World War, the service class was on the borderline between a caste and a class. Fascism was able to gain support not only from the bourgeoisie and the peasants but also from these castes, by flattering their caste pride and promising to maintain their position of superiority. According to different national conditions the totalitarian state supported itself in the first instance on the priestly caste (Spain, Dollfuss), on the military and bureaucratic caste (Germany), on the intellectuals (Italy), or on the new managers (Communism). Inflation, currency regulations, crises, war, and defeat have demolished these illusions.

This brief survey shows us that the stratum of those in service occupations has ceased to be merely a caste, and that in reality it has become a class—the service class just described—even if its caste spirit has not completely died out. The majority of its members have become in practice propertyless, since their inherited property is today either without value or at least socially insignificant. Thus the service class is closer to the rising working class in its life style, and at its boundary tends to merge with it. Nationalization and municipalization of powerful economic enterprises (railways, posts, water, light, and power supply) has established a bridge between the public services and private employment, between the situation of the worker, and that of the employee; and in crossing this bridge the caste spirit disappears. It is of the utmost importance for the eventual success of democratic socialism that these events should be closely studied and politically assessed.

On the other side, the trade union struggle has achieved for large sections of the working class a legal status which resembles that of officials; the trade union contract has become a kind of code of service. As a result of the technological revolution a great deal of work has come to involve the 'servicing' of machinery, and numerous occupational tasks increasingly require responsibility, as if they were positions in the public service.

Our analysis shows how superficially and carelessly many of those who claim to be Marxists approach the real study of class formation in society, and above all the continuous restructuring of the classes. It is abundantly clear that the factual substratum, the social basis, has been completely transformed in the past hundred years; that the working class as it appears (and scientifically was bound to appear) in Marx's *Capital* no longer exists. There remains only a narrow stratum of *déclassés* which is to serve the Communists as cannon-fodder in the struggle for power by the new caste which is to be presented to the world as its ruler; or as the dynamite to destroy a higher civilization which they do not understand and consequently hate—a civilization which will exclude all caste rule for ever, including that newly created by the Soviet bureaucracy and hierarchy.

FURTHER READING
*

Braverman, H. *Labour and Monopoly Capital* (1974)
Dahrendorf, Ralf *Class and Class Conflict in an Industrial Society* (1959), Part I
Hussain, Athar and Tribe, Keith *Marxism and the Agrarian Question* (2 vols. 1981)
Mallet, Serge *The New Working Class* (1975).
Mann, Michael *Consciousness and Action among the Western Working Class* (1973)
Poulantzas, Nicos *Classes in Contemporary Capitalism* (1975)

PART IV
*
The State and Politics

PRESENTATION

*

Marx and Engels conceived the state as a product of the division of society into classes and as a means by which a dominant class, the 'masters of the conditions of production', consolidates and maintains its rule over society. This conception is expressed most forcefully in a well-known passage of the *Communist Manifesto*: 'The executive of the modern state is but a committee for managing the common affairs of the whole bourgeoisie.' In other writings, however, Marx and Engels introduced some qualifications, notably concerning the 'relative independence' or 'autonomy' of the state in certain circumstances (see the following texts).

This question has become very prominent in recent Marxist studies, and diverse views have emerged. A fundamental problem is posed by the nature of the modern democratic state. Marx and Engels did not examine in any detail the implications of political democracy (which in any case had not advanced very far in their lifetimes), but Marx wrote of the Chartist campaign for universal suffrage that its success would be 'a far more socialistic measure than anything . . . on the Continent' and would mean 'the political supremacy of the working class.'[1] From the beginning of the twentieth century many Marxists, outstanding among them the Austro-Marxists, argued that socialism could, and would, be achieved through the institutions of political democracy, although its full development would require a considerable extension of democracy and the creation of new institutions such as workers' councils and other organs of self-government. Most recently, in the Euro-communist movement, there has been a vigorous rejection of the Leninist denigration of 'bourgeois democracy', and a reassertion of democratic ideas, summed up in the statement by Carrillo (1977, p. 105) that 'the political system established in Western Europe, based on representative political institutions . . . is in essentials valid . . . it is a question of making that system still more democratic'.

A second problem which has especially preoccupied Marxist

[1] Karl Marx, 'The Chartists', *New York Daily Tribune*, 25 Aug. 1852.

thinkers in recent decades concerns the growing power of the modern state associated with the massive scale of state intervention in the regulation of the economy and of social life as a whole. It has been generally accepted that this involves a change in the nature of capitalism (see Part VII below), but there has been considerable debate about how far this increasing power gives the state a much more independent role in social life and creates conditions for the emergence of a 'totalitarian state' (Hilferding 1941). Hence the question of the 'relative autonomy' of the state has become far more acute and controversial, but it remains unresolved, and no major theoretical work has yet provided a systematic analysis of the specific degree of independence of the state in diverse economic and social circumstances.

So far as the general analysis of modern political movements and trends is concerned, Marxist sociologists have had to confront a number of problems. The most crucial is no doubt that of the role of the working class, and working-class parties, in the changing social structure of Western capitalism (see Part III), and the future prospects of the whole socialist movement as traditionally conceived. But there are many related questions, particularly about the significance of recent social movements— the women's movement, the ecology movement, ethnic and nationalist movements, the student movement, the peace movement—and more broadly about the political influence of various social groups which are quite distinct from classes, and may have only tenuous links with any particular class.

Finally, there is the question of the state in socialist society, which Marx and Engels themselves did not examine in any thorough way. In his account of the Paris Commune,[2] which he described as 'the political form at last discovered under which to work out the emancipation of labour', Marx indicated some of its important features, including election of representatives by universal suffrage, and the proposed decentralization of government by the creation of local and district assemblies throughout France; and in a later text[3] he asked 'What changes will the state undergo in communist society? . . . What social functions will remain there which are analogous to the present functions of the state?', and went on to observe that 'this question can only be

[2] Karl Marx, *The Civil War in France* (1871).
[3] Karl Marx, *Critique of the Gotha Programme* (1875).

answered scientifically'. Engels, for his part, wrote of the 'withering away of the state', and adopted from Saint-Simon the idea of replacing the 'government of persons' by the 'administration of things' (see p. 130 below).

But this does not take us very far; and in particular the notion of 'administering things' is quite misleading if it does not recognize that all administration (for example, economic planning) necessarily takes place in a social context and cannot avoid some degree of 'government of persons'. At all events, in the existing socialist societies the state, far from withering away, has become immensely more powerful, and in many cases more oppressive. One inescapable task of Marxist sociology at the present time, therefore, is to investigate the conditions underlying the historical development of these highly centralized socialist states, and to consider afresh what kind of political institutions would be required in order to achieve, in reality, that general human emancipation which Marx saw as the content of socialism.

KARL MARX

*

The Division of Labour, Civil Society, and the State*

Finally, the division of labour offers us the first example of how, as long as man remains in natural society, that is, as long as a cleavage exists between the particular and the common interest, as long therefore as activity is not voluntarily, but naturally, divided, man's own act becomes an alien power opposed to him, which enslaves him instead of being controlled by him. For as soon as the division of labour begins, each man has a particular, exclusive sphere of activity which is forced upon him and from which he cannot escape. He is a hunter, a fisherman, a shepherd, or a critical critic, and must remain so if he does not want to lose his means of livelihood; whereas in communist society, where nobody has one exclusive sphere of activity but each can become accomplished in any branch he wishes, production as a whole is regulated by society, thus making it possible for me to do one thing today and another tomorrow, to hunt in the morning, fish in the afternoon, rear cattle in the evening, criticize after dinner, in accordance with my inclination, without ever becoming hunter, fisherman, herdsman, or critic.

This crystallization of social activity, this consolidation of what we ourselves produce into an objective power over us, growing out of our control, thwarting our expectations, bringing to naught our calculations, is one of the chief factors in historical development up to the present. It is precisely as a result of this contradiction between the interest of the individual and that of the community, that the latter takes an independent form as the *state*, divorced from the real interests of individual and community, and at the same time as an illusory community life, but always on the real basis of the bonds existing in every family and tribal aggregate, such as consanguinity, language, division of labour on a larger scale, and other interests. It arises especially, as

* *From The German Ideology* (1845–6), vol. i, pt. i, sect. A, 1. Translated by Tom Bottomore.

will be shown later, on the basis of social classes conditioned by the division of labour, which emerge in every aggregate of this kind, and of which one dominates all the others. It follows from this, that all the struggles within the state, the struggle between democracy, aristocracy, and monarchy, the struggle for the franchise, etc., are merely the illusory forms in which the real struggles of the different classes with each other are fought out (. . .); and further, that every class which is striving for domination—even when its domination, as is the case with the proletariat, involves the abolition of the old form of society as a whole, and of domination in general—must first seize political power in order to exhibit its interests as the general interest, as it is obliged to do at the very outset. Just because individuals seek *only* their particular interest, which for them does not coincide with their communal interest (and above all not with the 'general interest', the illusory form of community life), the communal interest will be established as something 'alien' to them and 'independent' of them, as a 'general' interest which is once again separate and distinct; or else they themselves will be obliged to move within this discord, as in democracy. On the other hand, the *practical* struggle of the enduring *real* special interests against communal, and illusory communal, interests also makes necessary *practical* intervention and restraint through the illusory 'general' interest in the form of the state. The social power, that is, the multiplied productive force which results from the combined efforts of individuals determined by the division of labour, appears to these individuals—since their co-operation itself is not voluntary, but imposed by nature—not as their own united power, but as an alien force existing outside themselves, of whose origin and destination they know nothing; which they can, therefore, no longer master. On the contrary, this force goes through its own specific series of phases and stages of development which are not only independent of human will and action, but control and direct them. (. . .)

Civil society comprises all the material intercourse of individuals within a specific stage of development of the productive forces. It comprises the whole commercial and industrial life of this stage, and in that respect extends beyond the state and the nation, although on the other hand, it must still assert itself externally as a nation and constitute itself internally as a state. The term 'civil society' came into use in the eighteenth century, when property relations had already worked themselves out of the communal life of antiquity and the Middle Ages. Civil society as such develops first with the bourgeoisie; the social organization developing directly out of production and intercourse, which forms at all times the basis of the state and of the rest of the idealist superstructure, has been described since then by this term.

KARL MARX
*
The State and Bureaucracy in France*

This executive power with its huge bureaucratic and military organization, its extensive and ingenious state machinery, its host of officials numbering half a million, besides an army of another half million, this terrible parasitic body which encompasses the body of French society like a caul and chokes all its pores, emerged in the days of the absolute monarchy, with the decline of the feudal system which it helped to accelerate. The seignorial privileges of the landowners and towns were transformed into so many attributes of the state power, the feudal dignitaries into paid officials, and the motley pattern of conflicting medieval plenary powers into the regulated plan of a state authority whose work is divided and centralized like that of a factory. The first French Revolution, with its task of breaking all separate local territorial, urban, and provincial powers in order to create the civic unity of the nation was obliged to develop what the absolute monarchy had begun, centralization, but at the same time the extent, the attributes, and the subordinate agents of governmental power. Napoleon perfected this state machinery. The Legitimist monarchy and the July monarchy added nothing but a greater division of labour, growing in the same measure as the division of labour in bourgeois society created new groups of interests, and hence new material for state administration. Every *common* interest was immediately detached from society and counterposed to it as a higher *general* interest, snatched away from the activity of the members of society themselves and made an

* From Karl Marx, *The Eighteenth Brumaire of Louis Bonaparte* (1852), sect. vii. Translated by Tom Bottomore.

object of government activity; from the bridge, the school, and the communal property of a village community, up to the railways, the national wealth, and the national university of France. Finally, in its struggle against the Revolution, the parliamentary republic was obliged to reinforce the resources and centralization of state power with repressive measures. All revolutions perfected this machine instead of smashing it. The parties which contended in turn for domination regarded the seizure of this huge state edifice as the principal spoils of the victor.

FRIEDRICH ENGELS
*
Origins and Development of the State*

The three main forms in which the state arises on the ruins of the gentile[1] constitution have been examined in detail above. Athens offers the purest, most classical form. Here the state arises directly and predominantly from the class oppositions which develop within gentile society itself. In Rome, gentile society becomes a closed aristocracy in the midst of a numerous *plebs*, which stands outside it, having duties but no rights; the victory of the *plebs* breaks up the old kinship constitution and erects on its ruins the state, into which both the gentile aristocracy and the *plebs* are soon completely absorbed. Finally, with the German victory over the Roman Empire, the state arises directly out of the conquest of large foreign territories, which the gentile constitution provides no means of ruling. But because this conquest involves neither a serious struggle with the original population, nor a more

* From Friedrich Engels, *The Origin of the Family, Private Property and the State* (1884), ch. ix. Translated by Tom Bottomore.

[1] Engels took the term 'gens' (and hence gentile society) from L. H. Morgan (1871, 1877) who, having discovered 'The identity of structures and function' in the Amerindian clan and in the *genos* or gens of the ancient Greeks and Romans, used the term 'gens' rather than 'clan', and referred to 'gentilitius' society rather than 'tribal society'. See further the discussion in Godelier (1977), pp. 72–6 [Eds.].

advanced division of labour; because the level of economic development of conquerors and conquered is almost the same, and the economic basis of society therefore remains as it was—so the gentile constitution is able to persist for many centuries in the changed territorial form of the mark[2] constitution and even to renew itself in an attenuated form in the later noble and patrician kin groups, and even in peasant kin groups as in Ditmarschen.[3]

The state, therefore, is by no means a power imposed upon society from outside; equally, it is not, as Hegel maintains, 'the reality of the moral idea', 'the image and reality of reason'. Rather, it is a product of society at a certain stage of development, the admission that this society has entangled itself in an insoluble self-contradiction and is split into irreconcilable oppositions which it is powerless to conjure away. But to prevent these oppositions—classes with conflicting economic interests—from consuming themselves and society in a fruitless struggle, a power apparently standing above society has become necessary in order to moderate the conflict and keep it within the bounds of 'order'. This power, which has emerged from society but places itself above it and becomes increasingly alienated from it, is the state.

In contrast to the old gentile organization the state is characterized in the first place by the classification of its members *on a territorial basis*. The old gentile associations, formed and held together by blood ties, had become inadequate, as we have seen, largely because they presupposed the attachment of the members to a particular region, and this had long ceased to be the case. The region remained but the people had become mobile. Hence the territorial division was taken as the starting point, and citizens were allowed to exercise their public rights and perform their duties where they settled, without regard to gens or tribe. This organization of members of a state according to their domicile is common to all states. Consequently, it seems quite

---

[2] Engels published as an appendix to the German edition of *Socialism: Utopian and Scientific* (1883) a study of the German *Mark*, as a community of villages in which there was some degree of communal ownership of land and communal regulation of public affairs, elements of which survived in certain parts of Germany until the beginning of the nineteenth century [Eds.].

[3] The first historian who had at least a rough conception of the nature of the gens was Niebuhr, and for that—though also for carrying over several misconceptions—he was indebted to his acquaintance with the Ditmarschen kin groups.

natural to us; but we have seen what hard and prolonged struggles were required before it could replace the old kinship organization in Athens and Rome.

The second characteristic of the state is the institution of a *public force* which no longer corresponds directly with the people's organization of themselves as an armed power. This special public force is necessary because a self-acting armed organization of the people has become impossible since the division into classes. Slaves are also part of the population; the 90,000 Athenian citizens constitute only a privileged class as against the 365,000 slaves. The people's army of the Athenian democracy confronted the slaves as an aristocratic public force which held them in check; but a police force was also necessary, as recounted above, to control the citizens. The public force exists in every state; it does not consist only of armed men but also of material appendages—prisons and coercive institutions of all kinds—which were unknown in gentile society. It may be quite insignificant, almost negligible, in societies where class oppositions have still not developed, and in remote areas, as has been the case at some times and places in the USA. But it becomes stronger to the extent that class oppositions within the state become more acute and neighbouring states become larger and more populous. It is enough to look at present-day Europe, where class struggle and competition in conquest have raised the public power to a level at which it threatens to devour the whole society and even the state itself.

In order to maintain this public power contributions from the citizens are necessary—*taxes*. These were completely unknown to gentile society. But we know more than enough about them today. With advancing civilization even taxes do not suffice; the state draws bills on the future, contracts loans—*state debts*. Old Europe can tell a tale about these too.

In possession of the public power and the right of taxation the officials now represent an organ of society standing *above* society. The free, willing respect which was accorded to the organs of the gentile constitution is not enough for them, even if they could have it; as the bearers of a power alienated from society they have to be given prestige by exceptional laws, thanks to which they enjoy a special sanctity and inviolability. The most miserable policeman of the civilized state has more 'authority' than all the organs of gentile society together; but the most powerful prince, and the greatest statesman or general of civilization, may well envy the lowliest gentile chief the unforced and unchallenged respect which he receives. For the one stands in the midst of society, while the other is obliged to pose as something outside and above it.

Since the state emerged from the need to keep class antagonisms in check, but also in the midst of the conflict between these classes, it is as a rule the state of the most powerful, economically dominant class, which also becomes, by this means, the politically dominant class and so acquires new instruments for subordinating and exploiting the oppressed. Thus the ancient state was above all the state of the slave-owners, for the purpose of holding down the slaves, just as the feudal state was the organ of the nobility for subordinating the peasant serfs and bondsmen, and the modern representative state is an instrument for the exploitation of wage labour by capital. There are, however, exceptional periods in which the contending classes are so nearly equal in strength that the state power, as an apparent mediator, temporarily acquires a certain independence in relation to both. Thus the absolute monarchy of the 17th and 18th centuries balanced the nobility and the bourgeoisie against each other, and the Bonapartism of the First, and especially the Second, Empire in France played off the proletariat against the bourgeoisie and the bourgeoisie against the proletariat. The latest achievement in this line, which turns rulers and ruled into equally comic figures, is the new German Empire of the Bismarckian nation; here capitalists and workers are balanced against each other and uniformly swindled for the benefit of the degenerate Prussian clod-hopping Junkers.

Further, in most historical states the rights accorded to citizens are graded according to property, and so express directly the fact that the state is an organization for the defence of the possessing class against the non-possessing class. This is already the case with the Athenian and Roman property classes, and similarly in the medieval feudal state, where the extent of political power corresponded with landownership. So also with the voting qualifications in modern representative states. This political recognition of property differences, however, is in no way essential. On the contrary, it indicates a low level of development of the state. The highest form of the state—the democratic republic—which becomes increasingly, in our modern social conditions, an inescapable necessity, in which alone the final decisive battle between bourgeoisie and proletariat can be fought out, no longer takes any official cognizance of property differences. Here the power of wealth is exercised indirectly, but all the more surely. On one side, in the form of direct corruption of officials, of which America provides the classical example; on the other side, in the form of an alliance between the government and the stock exchange, which is all the more easily accomplished the higher the state debt mounts and the more joint-stock companies concentrate in their hands not only transport but production itself, and find their own centre in the stock exchange. Besides America, the most recent French Republic provides a striking example, and honest little Switzerland has also made its contribution in this field. But the fact that a democratic republic is not essential for this fraternal link between government and stock exchange is shown not only by England but by the new German Empire, where it is hard to say whether Bismarck or the Bleichröder bank gained most from universal suffrage. As long as the oppressed class—in our case, therefore, the proletariat—is not yet ripe for self-emancipation, a majority of it will regard the existing social order as the only possible one and will be the tail, the extreme left wing, of the capitalist class. But to the extent that it matures towards self-emancipation, so it constitutes itself as its own party, and votes for its own representatives, not those of the capitalists. Thus universal suffrage is a gauge of the maturity of the working class. It cannot and will not ever be anything more in the present day state; but that is enough. On the day when the thermometer of universal suffrage indicates boiling-point among the workers, they as well as the capitalists will know where they stand.

The state, therefore, has not existed from all eternity. There have been societies which managed without it and had no idea of the state or state power. At a certain stage of economic development, which necessarily involved the division of society into classes, the state became a necessity because of this division. We are now rapidly approaching a stage in the development of production at which the existence of these classes has not only ceased to be a necessity but becomes a positive hindrance to production. They will decline with the same inevitability as they once arose. The state inevitably declines with them. The society which reorganizes production on the basis of free and equal association of the producers will put the whole state machine where it will then belong, in the museum of antiquities alongside the spinning wheel and the bronze axe.[4]

RALPH MILIBAND

*

The State and the Ruling Class*

The starting point of the Marxist theory of politics and the state is its categorical rejection of this view of the state as the trustee, instrument, or agent of 'society as a whole'. This rejection necessarily follows from the Marxist concept of society as class society. In class societies, the concept of 'society as a whole' and of the 'national interest' is clearly a mystification. There may be occasions and matters where the interests of all classes happen to coincide. But for the most part and in essence, these interests are fundamentally and irrevocably at odds, so that the state cannot possibly be their common trustee: the idea that it can is part of the ideological veil which a dominant class draws upon the reality of

---

[4] It is worth noting here the similar account of the decline of the state which Engels gave in *Socialism: Utopian and Scientific* (1880), sect. iii, because this passage contained some memorable phrases which later gave rise to much controversy: 'When [the state] finally becomes the real representative of the whole society it renders itself superfluous. As soon as there is no longer any social class to be held in subjection, as soon as class domination and the individual struggle for existence, based upon the prevailing anarchy of production, and the collisions and excesses which arise from them, are removed, there remains nothing to repress which would make necessary a special repressive force, a state. The first act through which the state really emerges as representative of the whole society—the seizure of the means of production in the name of society—is at the same time its last independent act as a state. The intervention of a state power in social relations becomes superfluous in one sphere after another and dies away of its own accord. The government of persons is replaced by the administration of things and the management of processes of production. The state is not "abolished"; *it withers away*' [Eds.].

* From Ralph Miliband, *Marxism and Politics* (Oxford: OUP, 1977), pp. 66–74, 83–4. Copyright © Ralph Miliband, 1977. Reprinted by permission of Oxford University Press.

class rule, so as to legitimate that rule in its own eyes as well as in the eyes of the subordinate classes.

In reality, so the Marxist argument has always been, the state is an essential means of class domination. It is not a neutral referee arbitrating between competing interests: it is inevitably a deeply engaged partisan. It is not 'above' class struggles but right in them. Its intervention in the affairs of society is crucial, constant and pervasive, and that intervention is closely conditioned by the most fundamental of the state's characteristics, namely that it is a means of class domination—ultimately the most important by far of any such means.

The most famous formulation of the Marxist view of the state occurs in the *Communist Manifesto*: 'The executive of the modern state is but a committee for managing the common affairs of the whole bourgeoisie.' But this is not nearly so simple and straightforward a formulation as it has commonly been interpreted to be. In fact, it presents, as do all Marxist formulations on the state, many problems which need careful probing. I do not mean by this that the general perspective is false: on the contrary, I think that it is closer to the political reality of class societies than any other perspective. But it is not a magic formula which renders the interpretation of that reality unproblematic. There is no such formula.

One immediate problem concerns the notion of 'ruling class' in its Marxist usage. In that usage, the 'ruling class' is so designated by virtue of the fact that it owns and controls a predominant part of the means of material and 'mental' production; *and* that it thereby controls, runs, dictates to, or is predominant in the state as well. But this *assumes* that class power is automatically translated into state power. In fact, there is no such automatic translation: the question of the relation between class power and state power constitutes a major problem, with many different facets. Even where that relation can be shown to be very close, a number of difficult questions remain to be answered, or at least explored. Not the least of these questions concerns the forms which the state assumes, and why it assumes different forms, and with what consequences.

But the very first thing that is needed is to realize that the relation between the 'ruling class' and the state is a problem, which cannot be assumed away. Indeed, the problem is implicit in the formulation from the *Manifesto* which I have quoted earlier. For the reference to 'the common affairs of the whole bourgeoisie' clearly implies that the bourgeoisie is a social totality made up of different and therefore potentially or actually conflicting elements, a point which, as was suggested in Chapter II, must be taken as axiomatic for all classes; while 'common affairs' implies the existence of particular ones as well. On this basis, there is an absolutely essential function of mediation and reconciliation to be performed by the state; or rather, it is the state which plays a major role in the performance of that function, the qualification being required because there are other institutions which help in its performance, for instance the parties of the bourgeoisie.

But if the state is to perform this mediating and reconciling function for what are, in effect, different elements or fractions of the bourgeoisie, which have different and conflicting interests, it clearly must have a certain degree of autonomy in relation to the 'ruling class'. In so far as that class is not monolithic, and it never is, it cannot act as a principal to an agent, and 'it' cannot simply use the state as 'its' instrument. In this context, there is no 'it', capable of issuing coherent instructions, least of all in highly complex, fragmented and 'old' societies, where a long process of historical development has brought to predominance a 'ruling class' which harbours many different interests and fractions. This is not to deny the acceptability, with various qualifications, of the term, but only to suggest that the relation of the 'ruling class' to the state is always and in all circumstances *bound* to be problematic.

The best way to proceed is to ask the simplest possible question, namely why, in Marxist terms, the state should be thought to be the 'instrument' of a 'ruling class'; and an answer to this question should also yield an answer to the question of the validity of the concept of 'ruling class' itself.

Marxists have, in effect, given three distinct answers to the question, none of which have however been adequately theorized.

The first of these has to do with the *personnel* of the state system, that is to say the fact that the people who are located in the commanding heights of the state, in the executive, administrative, judicial, repressive and legislative branches, have tended to belong to the same class or classes which have dominated the other strategic heights of the society, notably the economic and the cultural ones. Thus in contemporary capitalism, members of the bourgeoisie tend to predominate in the three main sectors of social life, the economic, the political and cultural/ideological—the political being understood here as referring mainly to the state apparatus, though the point would apply more widely. Where the people concerned, it is usually added, are not members of the bourgeoisie by social origin, they are later recruited into it by virtue of their education, connections, and way of life.

The assumption which is at work here is that a common social background and origin, education, connections, kinship, and friendship, a similar way of life, result in a cluster of common ideological and political positions and attitudes, common values and perspectives. There is no necessary unanimity of views among the people in question and there may be fairly deep differences between them on this or that issue. But these differences occur within a specific and fairly narrow conservative spectrum. This being the case, it is to be expected, so the argument goes, that those who run the state apparatus should, at the very least, be favourably disposed towards those who own and control the larger part of the means of economic activity, that they should be much better disposed towards them than towards any other interest or class, and that they should seek to serve the interests and purposes of the economically dominant class, the more so since those who run the state power are most likely to be persuaded that to serve these interests and purposes is also, broadly speaking, to serve the 'national interest' or the interests of 'the nation as a whole'.

This is a strong case, easily verifiable by a wealth of evidence. The bourgeois state *has* tended to be run by people very largely of the same class as the people who commanded the 'private sector' of the economic life of capitalist societies (and for that matter the 'public sector' as well). What I have elsewhere called the state élite *has* tended to share the ideological and political presumptions of the economically dominant class. And the state in capitalist society *has* tended to favour capitalist interests and capitalist enterprise—to put it thus is in fact to understate the bias.

Yet, strong though the case is, it is open to a number of very serious objections. These do not render the consideration of the nature of the state personnel irrelevant. Nor do they in the least affect the notion of the state as a class state. But they do suggest that the correlation which can be established in class terms between the state élite and the economically dominant class is not adequate to settle the issue.

One major objection is that there have been important and frequent exceptions to the general pattern of class correlation. These exceptions have occurred, so to speak, at both the upper and the lower levels of the social scale.

Britain provides the most notable example of the former case. Here, a landowning aristocracy continued for most of the nineteenth century to occupy what may accurately be described as an overwhelmingly preponderant place in the highest reaches of the state apparatus; while Britain was in the same period turning into the 'workshop of the world', the most advanced of all capitalist countries, with a large, solid, and economically powerful capitalist class. The phenomenon was of course familiar to Marx and Engels, who frequently remarked upon it. Thus in an article, 'The British Constitution', written in 1855, Marx observed that 'although the bourgeoisie, itself only the highest social stratum of the middle classes', had 'gained *political* recognition as the *ruling class*', 'this only happened on one condition; namely that the whole business of government in all its details—including even the executive branch of the legislature, that is, the actual making of laws in both Houses of Parliament— remained the guaranteed domain of the landed aristocracy'; and he went on to add that 'now, subjected to certain principles laid down by the bourgeoisie, the aristocracy (which enjoys exclusive power in the Cabinet, in Parliament, in the Civil Service, in the Army and Navy . . .) is being forced at this very moment to sign its own death warrant and to admit before the whole world that it is no longer destined to govern England'.

Of course, the aristocracy was then neither signing its death warrant nor admitting anything of the kind Marx suggested. But even if we leave this aside, there are here very large theoretical as well as empirical questions which remain unanswered concerning the relations of this new 'ruling class' to a landed aristocracy which, according to Marx, remained in complete charge of state power; and also concerning the ways in which the 'compromise' of which Marx also spoke in this article worked themselves out.

The same phenomenon of an aristocratic class exercising power 'on behalf' of capitalism occurred elsewhere, for instance in Germany, and was discussed by Marx and Engels, though in different terms. But it is obvious that this presents a problem for the thesis that the bias of the state is determined by the social class of its leading personnel.

The problem is at least as great where an important part of the state apparatus is in the hands of members of 'lower' classes. This too has frequently been the case in the history of capitalism. In all capitalist countries, members of the petty bourgeoisie, and increasingly of the working class as well, have made a successful career in the state service, often at the highest levels.[1] It may well be argued that many if not most of them have been 'absorbed' into the bourgeoisie precisely by virtue of their success; but the categorization is too wide and subjective to be convincing. In any case, there are telling instances where no such 'absorption' occurred: the most dramatic such instances are those of the Fascist dictators who held close to absolute power for substantial periods of time in Italy and Germany. This will need to be discussed later, but it may be noted here that whatever else may be said about them, it cannot be said about Hitler and Mussolini that they were in any meaningful way 'absorbed' into the German and Italian bourgeoisies and capitalist classes.

In other words, the class bias of the state is not determined, or at least not decisively and conclusively determined, by the social origins of its leading personnel. Nor for that matter has it always been the case that truly bourgeois-led states have necessarily pursued policies of which the capitalist class has approved. On the contrary, it has often been found that such states have been quite seriously at odds with their economically dominant classes. The classic example is that of Roosevelt after his election to the Presidency of the United States in 1932. In short, exclusive reliance on the social character of the state personnel is unhelpful—it creates as many problems as it solves.

The second answer which Marxists have tended to give to the question why the state should be thought to be the 'instrument' of

[1] In the *Prison Notebooks*, Gramsci asks: 'Does there exist, in a given country, a widespread social stratum in whose economic life and political self-assertion . . . the bureaucratic career, either civil or military, is a very important element?'; and he answered that 'in modern Europe, this stratum can be identified in the medium and small rural bourgeoisie, which is more or less numerous from one country to another', and which he saw as occasionally capable of 'laying down the law' to the ruling class (op. cit., pp. 212, 213).

a capitalist 'ruling class' has to do with the economic power which that class is able to wield by virtue of its ownership and control of economic and other resources, and of its strength and influence as a pressure group, in a broad meaning of the term.

There is much strength in this argument too, to which is added by virtue of the growth of economic giants as a characteristic feature of advanced capitalism. These powerful conglomerates are obviously bound to constitute a major reference point for governments; and many of them are multi-national, which means that important international considerations enter into the question. And there is in any case a vital international dimension injected into the process of governmental decision-making, and in the power which business is able to exercise in the shaping of that process, because governments have to take very carefully into account the attitudes of other and powerful capitalist governments and of a number of international institutions, agencies and associations, whose primary concern is the defence of the 'free enterprise' system.

Capitalist enterprise is undoubtedly the strongest 'pressure group' in capitalist society; and it is indeed able to command the attention of the state. But this is not the same as saying that the state is the 'instrument' of the capitalist class; and the pressure which business is able to apply upon the state is not in itself sufficient to explain the latter's actions and policies. There are complexities in the decision-making process which the notion of business as pressure group is too rough and unwieldy to explain. There may well be cases where that pressure is decisive. But there are others where it is not. Too great an emphasis upon this aspect of the matter leaves too much out of account.

In particular, it leaves out of account the third answer to the question posed earlier as to the nature of the state, namely a 'structural' dimension, of an objective and impersonal kind. In essence, the argument is simply that the state is the 'instrument' of the 'ruling class' because, *given its insertion in the capitalist mode of production*, it cannot be anything else. The question does not, on this view, depend on the personnel of the state, or on the pressure which the capitalist class is able to bring upon it: the nature of the state is here determined by the nature and requirements of the mode of production. There are 'structural constraints' which no government, whatever its complexion, wishes, and promises, can ignore or evade. A capitalist economy has its own 'rationality' to which any government and state must sooner or later submit, and usually sooner.

There is a great deal of strength in this 'structural' perspective, and it must in fact form an integral part of the Marxist view of the state, even though it too has never been adequately theorized. But it also has certain deficiencies which can easily turn into crippling weaknesses.

The strength of the 'structural' explanation is that it helps to understand why governments do act as they do—for instance why governments pledged to far-reaching reforms before reaching office, and indeed elected because they were so pledged, have more often than not failed to carry out more than at best a very small part of their reforming programme. This has often been attributed—not least by Marxists—to the personal failings of leaders, corruption, betrayal, the machinations of civil servants and bankers, or a combination of all these. Such explanations are not necessarily wrong, but they require backing up by the concept (and the fact) of 'structural constraints' which do beset any government working within the context of a particular mode of production.

The weakness of the case is that it makes it very easy to set up arbitrary limits to the possible. There *are* 'structural constraints'—but how constraining they are is a difficult question; and the temptation is to fall into what I have called a 'hyper-structuralist' trap, which deprives 'agents' of any freedom of choice and manoeuvre and turns them into the 'bearers' of objective forces which they are unable to affect. This perspective is but another form of determinism—which is alien to Marxism and in any case false, which is much more serious.[2] Governments can and do press against the 'structural constraints' by which they are beset. Yet, to recognize the existence and the importance of these constraints is also to point to the *limits of reform*, of which more later, and to make possible a strategy of change which attacks the mode of production that imposes the constraints.

Taken together, as they need to be, these three modes of explanation of the nature of the state—the character of its

---

[2] See R. Miliband, 'Poulantzas and the Capitalist State' in *New Left Review*, no. 82, Nov.–Dec. 1973, for a critical assessment of what I take to be an example of this type of determinism, namely N. Poulantzas, *Political Power and Social Classes* (London, 1973). For a reply, see N. Poulantzas, 'The Capitalist State: A reply to Miliband and Laclau', in *New Left Review*, no. 95, Jan.–Feb. 1976.

leading personnel, the pressures exercised by the economically dominant class, and the structural constraints imposed by the mode of production—constitute the Marxist answer to the question why the state should be considered as the 'instrument' of the 'ruling class'.

Yet, there is a powerful reason for rejecting this particular formulation as misleading. This is that, while the state does act, in Marxist terms, *on behalf* of the 'ruling class', it does not for the most part act *at its behest*. The state is indeed a class state, the state of the 'ruling class'. But it enjoys a high degree of autonomy and independence in the manner of its operation as a class state, and indeed *must* have that high degree of autonomy and independence if it is to act as a class state. The notion of the state as an 'instrument' does not fit this fact, and tends to obscure what has come to be seen as a crucial property of the state, namely its *relative autonomy* from the 'ruling class' and from civil society at large. (. . .)

The relative autonomy of the state was mainly acknowledged by Marx and Engels in connection with forms of state where the executive power dominated all other elements of the state system—for instance the Absolutist State, or the Bonapartist or Bismarckian one. Where Marx and Engels do acknowledge the relative autonomy of the state, they tend to do so in terms which sometimes exaggerate the extent of that autonomy. Later Marxist political thought, on the contrary, has usually had a strong bias towards the underestimation of the state's relative autonomy.

What this relative autonomy means has already been indicated: it simply consists in the degree of freedom which the state (normally meaning in this context the executive power) has in determining how best to serve what those who hold power conceive to be the 'national interest', and which in fact involves the service of the interests of the ruling class.

Quite clearly, this degree of freedom is in direct relation to the freedom which the executive power and the state in general enjoy *vis-à-vis* institutions (for instance parliamentary assemblies) and pressure groups which represent or speak for either the dominant class or the subordinate ones. In this sense, the relative autonomy of the state is greatest in regimes where the executive power is least constrained, either by other elements within the state system, or by various forces in civil society. Of such regimes,

Bonapartist France was the example with which Marx was most familiar. Its extreme version in capitalist society has been Fascism in Italy and Nazism in Germany, with many other less thorough examples in other capitalist societies. But it may be worth stressing again that *all* class states do enjoy *some* degree of autonomy, whatever their form and however 'representative' and 'democratic' they may be. The very notion of the state as an entity separate from civil society implies a certain distance between the two, a *relation* which implies a disjunction. The disjunction, which may be minimal or substantial, can only come to an end with the disappearance of the state itself, which in turn depends on the disappearance of class divisions and class struggles; and most likely a lot else as well.

CLAUS OFFE
*
Political Power in Late Capitalist Societies*

For political systems regarded as democratic, the question of ruling social groups or strata is not as self-evident as it may seem at first glance. *Democratic* rule should, after all, mean that within the social system as a whole *no* group is granted *a priori* particular power privileges. The historical importance of democratically constituted systems has been to render the politically dominant classes, strata, or groups functionally inoperative in a *ruling* capacity. Hence, the question from which an analysis of political authority must proceed should not, without qualification, imply the existence of a 'seat' or 'locus' of political authority within the social structure. Before proceeding in this manner, we must ascertain whether the question, so formulated, is indeed at all meaningful for democratic political systems. The first question, then, to which any systematic treatment of authority within

* From Claus Offe, 'Political Authority and Class Structures—An Analysis of Late Capitalist Societies', *International Journal of Sociology*, vol. ii, no. 1, pp. 73–81. The article first appeared in *Politikwissenschaft* (1969), pp. 155–89. Reprinted by permission of M. E. Sharpe, Inc., and Europäische Verlagsanstalt.

democratic political systems should address itself is not 'Who are the rulers and who are the ruled?' but 'Are we justified at all in applying *the category of ruling groups to democratic socio-political systems, and in so doing contesting their own postulate* that in these systems authority is vested in the constitution rather than in privileged power groups—an authority, moreover, that is beyond partisan politics and legitimated to like extent by all its citizens?'

Many theoretical writings on political sociology contain an implicit answer to this question and may be classified accordingly. The advocates of a *conflict theory* hold the view that even (and especially) late capitalist democratic political systems can be analysed properly only if they are viewed in light of an antagonism between ruler and ruled, between powerful oligarchies, classes, strata, and groups, on the one hand, and relatively powerless groups and classes whose interests are repressed in their expression and thwarted in their prosecution, on the other. This view, which is also shared by Marxist-oriented sociologists and political scientists, is directly countered by the other view, that of the *integration theorists*. They proceed in their analysis of political power from the postulate that society is integrated by a legitimate system of political institutions, which thereby neutralizes any structurally derived power advantages of particular groups, or at the very least ensures the maintenance of an unstable (pluralistic) equilibrium between competing power groups. In this view, a political system of democratic institutions carries out a series of functions which cannot be accounted for by the dominance of one group or the structural subordination of another. Political authority is not directly reducible to power derived from social structure. (. . .)

In political sociology, the controversy between the conflict and integration theories sets the tone for the problems that arise in attempting to establish a theoretical framework for the analysis of political authority. The spokesmen for the various perspectives on which such a framework might be based may be classified as conflict theorists in so far as their primary interest is in the social *basis* of power and the *distribution* of the prospects for influence within the social structure, or they may be classed as integration theorists if they are concerned principally with the *role played by a power potential*, regardless of its source or origin, in maintaining the stability of the total social system, i.e. to the extent that they place primary emphasis on the *employment* or *use* of institutions.

The question 'Which groups acquire power on the basis of what political conditions?' and 'What societal functions are served by the political sub-system as an agency for the use and application of power?' have up to now been treated separately, and to that extent inadequately, by political sociology. Neither position is able to provide solutions that would satisfy the dominant theoretical interest of its opponent. If one examines the answers of the 'distribution theorists' to the question of the political functions of the political exercise of authority, one usually encounters the circular argument that the functions consist essentially in maintaining particular power privileges. The distributive structure of prospects for political influence generate political functions which, in turn, serve mainly to perpetuate the existing pattern of power distribution. The functions generated by a structure are identical with it to the extent that they affect nothing but its continuity; the political system mediates the identity of the dominant class *as* dominant.

Even aside from the fact that this restrictive interpretation of the functions of political authority has never been empirically plausible, it immediately raises the theoretical problem at the basis of any explanation of total social change, i.e. of evolutionary and revolutionary *processes*. For any dynamic analysis to be effective, it must be able to deal with *unanticipated consequences* as something more than residual phenomena, that is, as a category of events of central importance to the system. In other words, structure and function must not exhibit perfect congruence.

A systems analysis of political authority from the integrative viewpoint runs up against a similar dilemma. In this view, the political system generates and distributes a certain type of functionally required facilities for the counter-system.

Within this framework, the question of the concrete political distribution of prospects for political influence has meaning only with reference to the criterion of effective integrative performance. The ruling group is always the one which demonstrates the greatest actual or prospective capacity for getting things done in the collective interest of society. Thus, power is manifested no longer merely as the privilege of this group, but as a functionally efficient and arbitrarily expandable product (like money) of total social systems, which the latter generate and apply for the regulation and control of their equilibrium conditions. In this view, then, the conflict theorists' problem of the structural distribution of prospects for political influence among structurally determined social groups becomes meaningless. Privileges falling to particular social groups can never be regarded as anything more than side effects of a regulatory process bearing on the whole of society. The integration theorists are as inclined to deny the relevance of the structural dimensions of the distribution of power as the conflict theorists are to see a political system with self-sufficient, i.e. structurally independent, functions at an analytical level.

Moreover, the ideological implications of the integration approach are quite obvious: by viewing political action and decision as a process that generates regulatory facilities and promotes so-called social goals, assumed to be independent of particular interests, it renders the relationships existing between social privileges and political authority inaccessible to analysis. The repressive factor inherent in any organization of political power evaporates, and accordingly the critical perspective necessary for emancipation from this repression becomes irrelevant.

Nonetheless, this perspective has a certain cogency. The comprehensive state regulation of all the processes vital to society—a feature which distinguishes late capitalist societies from their original bourgeois forms—is certainly brought into better focus and given more explicit treatment by the integration approach than by the competing conflict model. If one approaches the empirical power relations obtaining between societal interest groups as an unstable equilibrium kept in balance by the state apparatus, rather than as a pre-political, natural phenomenon, then the domain of 'civil society', with its semblance of autonomy, must be described in terms that bring out its politically mediated nature. In an era of comprehensive state intervention, one can no longer reasonably speak of 'spheres free of state interference' that constitute the 'material base' of the 'political superstructure'; an all-pervasive state regulation of social and economic processes is certainly a better description of today's order. Under these conditions, the official mechanisms sustaining the relationship between the state and society, such as subsidies, co-option, delegation, licensing, etc., reflect but subtle gradations of political control; as such, they help to maintain the fiction of a clear separation between state and society that has in fact become almost irrelevant.

This perspective, which any description of late capitalist welfare and intervention states must take into account, is more compatible with the integration approach, inasmuch as it sees political authority in the light of system-stabilizing performances rather than as an unmediated phenomenon representing particular interests, as in the opposing conflict theory.

The opposite perspective is spelled out by the theoretical conceptions of the conflict theory. Political authority appears merely as the specific form in which contradictory social interests are articulated, it being immaterial whether these interests are the expression of class structure, interest groups, political parties, power élites, or voters' behaviour. To the proponents of this view, the institutions of the political system are important mainly in their *instrumental* function—that is, as a means of power with whose help the pre-political socio-economic structure underlying these interests can be fortified and perpetuated.

Thus, our attempt to provide a theoretical framework for the analysis of the structures of authority in state-regulated capitalist systems has given rise to a dilemma. As we can no longer regard the system of political authority as a mere reflex or subsidiary organization for securing social interests, we are forced to abandon the traditional approach, which sought to reconstruct the political system and its functions from the elements of political economy. On the other hand, by so doing, we run the risk of losing sight of the fundamental authoritarian nature inherent in the organization of political power in late capitalist societies; we would then end up with a political system detached from its autonomous substrata, and as such absolved of any imputed repressed functions. This would amount to the conception that authority is tantamount to the mere temporary constitutional exercise of power and, moreover, is neutralized within a formal democratic system of institutions that conceal the shifting, pluralistic exercise of power. We can resolve this dilemma only by charting out the concrete mechanisms mediating between economics and politics—mechanisms which, on the one hand have 'politicized' social commodity-exchange down to the last detail, but which, on the other hand, have in no sense neutralized the *politicized* economy as the ultimate regulator in the functioning of political institutions.

The political constitution of liberal capitalist societies ensures the dominance of the pre-political interests of a ruling class in two ways which allow the political institutions themselves to be analysed in class terms. First, by means of the processes of political recruitment and consensus formation in a liberal constitutional system of a 'voluntary' party type, the bourgeoisie was able to utilize its ideology, political maxims, and value systems to create the preponderance that enabled it to bring the state's foreign policy, financial policy, and social policy confidently into line with its own interests. The second and most important means of co-ordinating the state apparatus with dominant capitalist interests was through mechanisms that *strictly delimited spheres of activity beyond state authority* and permitted their undisturbed use by economic persons. The restriction of state action to the functions of maintaining public order (military, the courts, police), which were carried out under a strictly neutral financial policy, created the conditions for private capital accumulation. Indeed, the bourgeois state confirmed its class nature precisely through the material limits it imposed on its authority. Any attempts to extend it would have been met with overwhelming resistance from legal institutions within the private sphere, e.g. property, the family, and contractual law.

Today we must face the fact that the defensive function of these legal institutions is no longer able in any serious way to limit the sphere of activities of the political system. Within the total system of late capitalist societies, social processes will almost without exception no longer take place beyond politics; on the contrary, they are regulated and sustained by permanent political intervention.

Thus, if we must assume that such a sphere of commodity exchange exempt from state power is no longer institutionally guaranteed in late capitalist welfare and social states, but that, on the contrary, an all-pervasive system of mechanisms for state intervention has been established, then the question of the concrete channels mediating social interests and political authority must be posed in a new way, namely: *What mechanisms ensure the dominant influence of social interests on the functioning of the political system even though these interests are no longer able to assert their former independence of the system within a free sphere beyond state power?*

In general, the repressive character of a political system—that is, those of its aspects serving to strengthen authority—is measurable in terms of whether it exempts *de facto* certain *spheres of action*, corresponding to the interests of particular groups, from the use of public force, so that these areas become sanctioned as natural and inviolable. The authoritarian character of a system of political institutions, on the other hand, is reflected in whether it accords equal *prospects for political consideration* to all the various classes of mutually incompatible social *interests, needs, and claims*, or whether these prospects are distorted or biased in some specific direction. Such a general concept of dominance or authority enables us to see beyond the momentary prospects of particular power groups and to evaluate the repressive factor intrinsic to the *institutions* distributing these prospects.

In contrast to late capitalist society, in the political structures of the liberal capitalist society the economic system was institutionalized as a domain beyond the authority of the state, and the economically dominant class possessed a *de facto* monopoly in the control of political decisions. Authority, then, was so structured that both *the limits defining the range of action of the political system, as well as those defining the prospects for the political articulation of needs, were commensurate with economically drawn class lines*. Only under such conditions can *political economy* provide the key to an analysis of the overall structures of dominance. Under the conditions of late capitalism, any attempt to explain the political organization of power through the categories of political economy becomes implausible. That is to say, *both* sides of the politically represented class relationship become problematic under the institutional conditions of late capitalist, democratically constituted societies. On the one hand, the boundaries of the exclusive domain of private interests (i.e. not merely in the sense of licensed autonomy or decentralized administration of tasks) are no longer delimitable in view of the state's universal right to intervene. On the other hand, it becomes difficult to discover, within the consensus formation of pluralistically organized interests and of universal suffrage, the institutional barriers that prevent specific interest groups within the society from participation in consensus formation. The late capitalist welfare state bases its legitimacy on the postulate of a universal participation in consensus formation and on the unbiased opportunity for all classes to utilize the state's services and to benefit from its regulatory acts of intervention.

RUDOLF HILFERDING

*

Democracy and the Working Class*

I have always been astonished by the assertion, which is still occasionally uttered, that democracy has been an affair of the bourgeoisie. Such a claim reveals ignorance of the history of democracy and a shallow intellectual attempt to derive this history from the writings of a handful of theorists. The truth is that there has been no political struggle more acute than that which the proletariat has waged against the bourgeoisie over democracy. If we do not realize that this struggle is one of the great deeds of the working-class struggle, and that it is false and misleading to speak, in a historical sense, of 'bourgeois democracy', then we are rejecting the entire socialist past from the time of Marx's famous comment that it is a matter of elevating the working class into a political party. Democracy has been *our* affair. We have had to wrest it from the bourgeoisie in a stubborn struggle. Remember the struggles for the right to vote. How much proletarian blood was shed in order to secure equal suffrage.

However, the term bourgeois democracy is not only wrong historically, but also from the standpoint of social analysis. For democracy signifies an entirely different method of constituting the will of the state. In the authoritarian state we confronted, in addition to the body of citizens constituted by elections and giving expression to its own political will, a number of powerful social organizations. I cannot pursue that point here; it is enough to say briefly that in fact, in all matters of vital importance, the will of the Reichstag counted for next to nothing as against that of the military high command, the high officials and the monarch. At the present time the will of the state is formed simply from the political will of individuals. The Reichstag is no longer confronted by exclusive, closed organizations of the rulers, and the latter must now address themselves to the citizens of the state and have their authority confirmed by a majority at repeated intervals, in an intellectual contest with us. If it is not confirmed,

---

* From Rudolf Hilferding, 'Die Aufgaben der Sozialdemokratie in der Republik' (1927). Translated by Patrick Goode. Published by permission of Dr Peter Milford.

then *as far as democracy is concerned* their rule is at an end.

But what if the rulers do not respect democracy? Is that a problem for us? Does it not go without saying, not only for every Social Democrat but—and I make this point deliberately—for every republican, that the moment there is any attempt to destroy the foundations of democracy all means are to be used in protecting them? What is involved here is the question of the use of force. After the experience of Germany in 1918, and especially of Russia, the use of force in the class struggle (and I refer to a real use of force: cutting, stabbing, and shooting) can only mean, not a transient putsch, but a long drawn out, very bitter and extremely bloody *civil war*. When the foundations of democracy are destroyed we are *on the defensive* and have no choice. Then we are obliged to employ all means. But no socialist—and I say this precisely from a socialist standpoint—will assert: Socialism does not attract me if I cannot use force in order to bring it about. Let me quote Otto Bauer: 'We will not do that because we know that there is no greater obstacle to the realization of socialism than civil war, and because our situation as socialists would be exceedingly difficult if proletarian state power issued from a civil war.' (. . .)

Democracy is a concern of the working class from a *sociological* as well as a historical perspective. Once again, it is completely unhistorical to suppose that democracy in the ancient world, early Italian democracy, and our modern form of democracy are comparable. Modern democracy exists only where strong proletarian organizations imbued with political consciousness support it; otherwise it comes to grief. Look at the South American states. Splendid constitutions, democracy installed, but no proletarian organizations, an economy run by democracy is only possible where a strong, class-conscious working class upholds it. The same goes for the countries of the East.

The expression *formal democracy* is equally mistaken, because it involves a misconception of the intimate connection between politics and the social consequences of politics. Democracy means that a different distribution of political power has been fully, or so far as possible, accomplished. This naturally means that the social consequences will be different, and the will of the state will be formed in a socially different way. The separation between politics and its social effects can be made in a theoretical and abstract way in writing, but in political reality this separation is quite artificial. From this standpoint too democracy is a fundamental concern of the proletariat. It is entirely wrong to describe democracy as formal. It is of the greatest *intrinsic* significance for the fate of every single worker.

There are people running about who shout: 'Beware of democratic illusions!' When Marx in his early writings—before the *Communist Manifesto*—made the point that political emancipation is not enough and that it must be complemented by human emancipation (what we now call social emancipation), that was of the greatest significance in relation to the bourgeois democrats of 1848, and was a very important educational task accomplished by Marx to protect the workers, who supported these democrats, from illusions. But is it not a feeble intellectual notion that we need constantly to warn the worker, who experiences in the factory every day for eight or ten hours in his own person the fact that political emancipation is not yet equivalent to social emancipation, against the illusions of bourgeois democracy? That is a piece of intellectual childishness which we should assail. My own view is quite different. The real danger, and one which has unfortunately not always remained only a danger, is that there have been sections of the working class and even the whole working class in some countries, which have *failed to recognize the importance of freedom and of democracy*. We have always been indignant, quite rightly, about the abandonment by the bourgeoisie of its liberal principles. But I have become more cautious in my own criticism since I saw how Mussolini came to power in Italy because the Italian proletariat did not recognize the value of freedom and democracy.

What applies to the South applies doubly and trebly to the East. The most depressing hours of my life in the party were those when I had to struggle in the Independent Social Democratic party (USPD) against the adherents of the Moscow 'Twenty-one Points'.[1] Many workers did not understand just what they were surrendering when they submitted to these twenty-one dictatorial conditions, which affected not only political life in general but even their own party. Since then we have discovered just what a misfortune Bolshevism has been. Whether the effects of Bolshevism have been reactionary or revolutionary will be for

[1] The 'Twenty-one Points', worked out by Lenin and Zinoviev, and adopted at the Second Congress of the Communist International (1920), laid down the conditions on which parties would be admitted to membership [Eds.].

history to judge in due course. But there is no doubt that for us Germans, and for the whole of Central Europe, the victory of the Bolsheviks *before* the victory of the democratic revolution in Germany was a great misfortune. If at that time we had *all* adhered firmly to democracy we would have overcome much more rapidly the division in the working class, and would have been able to achieve quite different, much greater successes, than was actually the case, simply because a part of the working class, by failing to recognize the importance of political rights, fought against its own front. If there are illusions to be destroyed, they are today no longer those which Marx destroyed in 1848. That is a completely absurd piece of pedantry. We must destroy those illusions which are dangerous *today*, and that means these anti-democratic illusions.

TOM BOTTOMORE
*
*The Working-Class Movement**

It is at this point that the second contradiction—between working class and bourgeoisie—which differs from the first in expressing an opposition of interests rather than an incompatibility of structures, assumes great importance. The demise of capitalism can only be the consequence of a political struggle, and it is the course of this struggle between classes, in the conditions created by the development of the capitalist mode of production, which we have now to examine. Marx's own theory of the class struggle was hardly more than sketched, in occasional passages which are scattered throughout his writings, and it was never presented in a systematic way. Moreover, the material upon which he had to work was that of the early stages of industrial capitalism and of the working-class political movement, so that in any case it would be essential to review his theory in the light of subsequent historical experience. (. . .)

---

* From Tom Bottomore, *Political Sociology* (London: Hutchinson University Library, 1979), pp. 30–6. Reprinted by permission of Hutchinson Publishing Group Ltd.

There are two broad sets of problems to be considered: first, the political impact of the working-class movement in capitalist societies as they have developed since the late nineteenth century; and second, the political systems that have emerged from revolutions carried out under the banner of Marxism as 'proletarian revolutions', in Russia, China, and other countries. As to the first question, it is evident that the working-class movement has had a profound influence upon the extension of the suffrage and the creation of mass parties (. . .) and hence upon the establishment of a democratic political regime as it now exists in the advanced capitalist countries. Furthermore, the pressure of working-class parties and trade unions has helped to produce a much more substantial intervention by the state in the economy, and this situation, although it can be interpreted from one aspect as the emergence of a new type of 'organized' or 'managed' capitalism, does also constitute a degree of protection of working-class interests against the power of capital through the general regulation of economic activity and the provision of an extensive network of social services, however imperfectly this may be done.

But it is equally obvious that the working class in the advanced capitalist societies has not been, for the most part, revolutionary in its outlook and action; least of all in the most advanced country, the USA. It is not that revolutionary movements have been absent (. . .) but that they have failed to elicit sustained and effective support from any large part of the working class. Of course, there have been historical fluctuations in revolutionary activity, as well as considerable variations between societies—with revolutionary parties having a much greater influence in France and Italy than in the rest of Western and Northern Europe or in North America—but the predominant style of working-class politics everywhere has been reformist, directed toward a gradual attrition of the unregulated market economy. There has not occurred that stark polarization and revolutionary confrontation of the two principal classes—bourgeoisie and proletariat—that Marx, at least in some parts of his analysis, seemed to anticipate.

How is this historical development to be explained? In the first place, perhaps, as many social thinkers from Bernstein to Schumpeter and beyond have suggested, by the economic successes of capitalism. Although there have been periods of economic stagnation or crisis, including the exceptionally severe crisis of the 1930s (which itself failed to engender large-scale revolutionary movements in most of the capitalist countries), the general tendency of capitalism has been to promote a continuous, and sometimes rapid, improvement in material levels of living, in which a large part of the working class, if not the whole class, has shared. This factor of material prosperity, which was already cited by Sombart at the beginning of this century as a partial explanation of the absence of any large-scale socialist movement in the USA,[1] acquired particularly great importance in the period following the Second World War, when economic growth took place more rapidly than ever before, and the question could be posed whether the United States did not simply show to capitalist Europe the image of its own future—a future that would be characterized by a decline of the socialist movement, and indeed of all ideological revolutionary parties and movements.[2]

This increasing prosperity is not, however, the only factor that can be adduced to explain the absence of successful revolutionary movements, or of such movements altogether, in Western capitalist societies. The class struggle, it may be argued, has been moderated by the incorporation of the working class into a modified and reformed capitalism through the extension of political, social, and economic rights, and the elaboration of a complex structure of contestation, bargaining and compromise within the existing form of society. The class struggle is further moderated, and turned increasingly into reformist channels, by changes in the nature of the class structure, and notably by the growth of the middle classes.

Working-class action in the economic sphere, through the trade unions, necessarily takes place in the context of a factual interdependence between employers and workers, and this interdependence is reinforced by the institutionalization of industrial conflict. Durkheim, in his discussion of the 'abnormal

---

[1] Sombart (1906).
[2] The idea was formulated by S. M. Lipset in an essay on 'The Changing Class Structure and Contemporary European Politics', in Graubard (1964). During the 1950s there was widespread debate among social scientists about this 'embourgeoisement' of the Western working class and about the 'end of ideology'; for two different analyses of some of the main issues see Goldthorpe *et al.* (1969), and Marcuse (1964).

forms' of the division of labour,³ long ago drew attention to what he regarded as a condition of 'anomie' in the sphere of production, characterized by the absence of a body of rules governing the relations between different social functions—above all, between labour and capital—and saw as both probable and desirable a growing normative regulation of industrial relations. This has in fact occurred in the advanced capitalist countries, and one main consequence has been to limit industrial conflict to economic issues, as against larger issues of the control of the enterprise, and to bring about a substantial degree of integration of workers into the existing mode of production.⁴ Hence it might be argued (as Marcuse and others have done) that a large part of the Western working class has been effectively incorporated into the economy and society of advanced capitalism, not only in the sphere of consumption, as a result of increasing prosperity, but also in the sphere of production, through the increasingly elaborate regulation of industrial relations by law and custom, and through the apparent technological imperatives of a high productivity, high consumption society.⁵ The outcome may thus be seen as a situation in which there is considerable trade-union militancy with respect to wages, hours of work, and related issues, but relatively little expression of class-consciousness in the broader, more revolutionary sense of a profound awareness or conviction of living in a society the nature of which is predominantly determined by class relations, and of being engaged in a continuing intense struggle to establish an alternative form of society.

There can be little doubt either that the development of working-class consciousness has been profoundly affected by changes in the class structure. The emergence and growth of new middle-class strata in the Western capitalist societies is a phenomenon which has raised questions for both Marxist and non-Marxist social scientists since the end of the nineteenth century. It was one important consideration in Bernstein's 'revision' of Marxist theory,⁶ and the problem was later analysed

---

³ Durkheim (1893), pt. iii.
⁴ See the discussion of this question, and some related issues, by Michael Mann (1973).
⁵ This last feature is particularly emphasized by Marcuse, op. cit.
⁶ See Bernstein (1899) and the study of Bernstein's views in Gay, *Democratic Socialism* (1952).

more fully by the Austro-Marxist thinkers, who came to recognize the political significance of the growing complexity of the class structure,[7] and in particular the influence of what Karl Renner called the 'service class'.[8] Among non-Marxist writers, especially during the 1950s, the expansion of these middle strata was often interpreted as marking the advent of middle-class societies in which there would be no fundamental cleavages or conflicts.[9] It is clear in any case that the growing size of the middle class must change fundamentally the image of capitalist society as one in which class antagonisms are simplified, and 'society as a whole is more and more splitting up into two great hostile camps, into two great classes directly facing each other— bourgeoisie and proletariat', which Marx and Engels depicted in the *Communist Manifesto* and which Marxists and other socialists generally accepted without much questioning until the early years of the twentieth century. The need arises to analyse and evaluate the probable political attitudes and actions of various middle-class groups in relation to the working class and the socialist movement, to right-wing parties and movements (more particularly, in the context of European politics in the 1920s and 1930s, and of Latin American politics in recent years, to fascist movements), and to diverse types of independent politics in the form of liberal or populist parties. A number of different answers has been given to the questions thus raised. One, already mentioned, envisages a gradual consolidation of a new kind of middle-class, post-industrial society,[10] based upon advanced technology, a mixed economy, and a broad consensus of opinion

---

[7] See especially the discussion in Rudolf Hilferding's last, unfinished work, *Das historische Problem* (first published, with an introduction, by Benedikt Kautsky in *Zeitschrift für Politik* (new series), vol. i (1954), pp. 293–324). Partly translated in Bottomore (1981), pp. 125–37.

[8] For Renner's study see Bottomore and Goode (eds.) (1978), pp. 249–52.

[9] See, among others, the writings of Raymond Aron, Daniel Bell, and S. M. Lipset. The idea is well expressed in Aron's observation that ' . . . experience in most of the developed countries suggests that semi-peaceful competition is gradually taking the place of the so-called deadly struggle in which one class was supposed to eliminate the other' (Aron (1968), p. 15). I have discussed the diverse interpretations of the changing class structure more fully in two essays, 'In Search of a Proletariat' and 'Class and Politics in Western Europe', reprinted in Bottomore (1975), chs. 6 and 8.

[10] Daniel Bell, *The Coming of Post-Industrial Society* (Basic Books, New York, 1973).

about social and political goals, which would be peaceful, liberal, and in a certain sense 'classless'.[11] Another foresees a more conservative role for the middle class, expressing itself in active opposition to socialism as a process of increasing public ownership or control of industry and expanding welfare services, and in a reassertion of the desirability of a more *laissez-faire* type of economy.[12]

A third kind of analysis conceives some sections of the middle class (technicians, managers, engineers, professional employees in the public service and in private industry) either as constituting an important part of a 'new working class' which is likely to participate in its own way in a refashioned socialist movement,[13] or as forming one element—alongside the old industrial working class—in a new class, which is becoming involved in a new type of struggle, directed against those who control the institutions of economic and political decision making, and who reduce it to a condition, not of misery or oppression, but of restricted and dependent participation in the major public affairs of society.[14]

One common theme in these varied accounts of the changing economy and class structure of the advanced capitalist societies is a questioning of the pre-eminent role of the industrial working class in bringing about a fundamental, revolutionary transformation of society from capitalism to socialism. But there is now also a wider questioning of the whole conception of a transition to socialism.[15] What J. A. Schumpeter, in *Capitalism, Socialism and Democracy*, called the 'march into socialism' seems to have slackened its pace; and socialism, which appeared in the nineteenth century as the ideal image of an alternative society, providing an indispensable unifying element in working-class consciousness, has become in the twentieth century a problematic reality. It is not simply that socialist societies in Eastern

---

[11] See the illuminating discussion of 'non-egalitarian classlessness' in Stanislaw Ossowski (1963), ch. vii.

[12] This kind of analysis is best exemplified in such works as Friedmann (1969) and Hayek (1973–8). It was also formulated, in a more qualified way, in the course of a critical assessment of socialist policies, by Schumpeter (1976).

[13] Mallet (1975).

[14] Touraine (1971).

[15] See the discussions in Parekh (1975), Kolakowski and Hampshire (1974), and Svetozar Stojanović, *Between Ideals and Reality: A Critique of Socialism and its Future* (1973).

Europe and elsewhere in the world are characterized, to a greater or lesser extent, by economic backwardness and political authoritarianism, and consequently have little appeal as models for the future development of any advanced industrial society, but that the democratic socialism of social democratic and labour parties in the capitalist world, despite its real achievements in improving the conditions of life of the working class, has come to be more critically judged as tending to promote an excessive centralization of decision-making, growth of bureaucracy, and regulation of the lives of individuals, and has lost something of the persuasive character it once had as a movement aiming to create a new civilization. Such changes in social and political thought clearly have important consequences for the character and goals of political action in the late twentieth century, and their effects are reinforced by the emergence of new problems and new movements—concerned with such issues as the environment and the use of natural resources, the subordination of women—which arguably have little connection with class politics; as well as by the renewed vigour of ethnic and national consciousness, expressed in independence movements of various kinds.

Elsewhere[16] I have distinguished, and tried to assess in relation to more traditional class politics, four new styles of political action: that of élites committed to rationality in production and administration, who justify their dominance by the benefits of sustained economic growth; that of various movements, notably the student movement of the 1960s, which attack technocracy and bureaucracy, and assert the counter-values of 'participation' and 'community'; that of regional and nationalist movements which also proclaim the value of community, founded upon a cultural identity; and that of various supranational movements—for example, Pan-Africanism, or the European Community—which attempt to organize whole regions of the world in terms of historical traditions, economic interests and cultural similarities. To speak of these 'new styles' of politics is not, of course, to say that political struggles between classes have ceased, but only that they may now be modified by other kinds of political action and be less predominant in political life as a whole, or to argue that the nature, aims, and strategies of the principal classes have changed substantially, as Mallet and

[16] Bottomore (1975), pp. 129–31.

Touraine have done. Undoubtedly, the two interrelated movements—the democratic movement and the labour movement—which developed so vigorously in the nineteenth century, continue to have a major influence in politics. However, the relation between them has changed during the present century, in a way which is also relevant to the character of more recent movements. In large measure, the nineteenth-century labour movement could be regarded—and regarded itself—as a continuation of the democratic movement, this continuity being expressed even in the name 'social democratic' which was generally adopted by the political parties of the working class. The idea which lay behind the use of the term 'social democracy' was that the working-class movement would not only complete the process of achieving political democracy by establishing universal and equal suffrage (and this itself required a long struggle), but would also extend democracy into other areas of social life, in particular through a democratic control of the economy, and would thus create new democratic institutions. One of the most important of these institutions was the workers' council, which made its appearance in the early part of this century as a method of establishing direct working-class control of production, in opposition to both capitalist ownership and centralized state control. The 'council movement' was especially vigorous, and was widely debated among socialists in the years immediately preceding and following the First World War;[17] and in the past two decades it has again aroused growing interest, as a result of the experience of workers' self-management in Yugoslavia (and some tentative steps in that direction in other East European countries), and of the formulation of ideas about 'participatory democracy' in the new social movements of the 1960s.[18]

---

[17] See Pribicevic (1959), and the study by Karl Renner, 'Democracy and the Council System', part of which is translated in Bottomore and Goode (1978), pp. 187–201.

[18] For general accounts of the Yugoslav experience, set in a wider context, see Blumberg (1968) and Broekmeyer (1970).

FURTHER READING

*

Bauer, Otto 'Fascism', in Tom Bottomore and Patrick Goode (eds.), *Austro-Marxism* (1978), pp. 167–86

Cummins, Ian *Marx, Engels and National Movements* (1980)

Gramsci, Antonio 'The Modern Prince' and 'State and Civil Society', in Quintin Hoare and Geoffrey Nowell Smith (eds.), *Selections from the Prison Notebooks* (1971), pp. 125–204, 210–76

Miliband, Ralph *The State in Capitalist Society* (1969)

Poulantzas, Nicos *Political Power and Social Classes* (1973)

PART V
*
Culture and Ideology

PRESENTATION

*

In the *German Ideology* (1845-6) Marx and Engels sketched the outlines of a theory of culture and ideology, but in their later work they did not return to the subject in any systematic way, and it was their Marxist followers, of various persuasions, who extended the theory to studies of aesthetics, law, language, the history and sociology of science, political ideologies, and other cultural phenomena. One of the first to do so was Mehring who, in *Die Lessing-Legende* (1893) and in his essays of literary criticism, laid the basis of a Marxist sociology of literature and of the history of ideas, while at the same time he attempted to construct a 'scientific aesthetic'.[1] Somewhat later, the Austro-Marxists made major contributions to the study of culture, particularly in two works—still unsurpassed today—by Bauer (1907) on nationalism and by Renner (1904) on law.

But it was from the 1920s that cultural studies began to acquire major importance in Marxist theory, with the writings of Lukács on class-consciousness, realism in literature, and general aesthetic theory, the work of the Frankfurt School (especially Adorno and Benjamin), and Gramsci's studies of the formation and social functions of intellectuals (though these became widely known only in the 1950s). During this period there was gradually formed, in the view of some scholars, a distinctive 'Western Marxism', characterized above all by a preoccupation with cultural questions, rather than economic or political problems, and to some extent with themes which were not in any obvious way Marxist at all; for example, with Simmel's 'tragedy of culture', or with the 'human condition' in an existentialist sense.[2]

This orientation became especially pronounced in the works of

---

[1] Mehring's very 'economistic' conception of historical materialism, however, provoked an implied criticism from Engels who, in his letter acknowledging receipt of *Die Lessing-Legende* (Engels to Mehring, 14 July 1893), declared once again (cf. also pp. 47 and 48 above) that he and Marx, and some of their followers, had unduly neglected the form in which ideologies develop, as against their content, and hence had provided opportunities for misunderstandings and distortions: for example, in respect of the historical influence of ideologies.

[2] See Arato and Breines (1979).

the later Frankfurt School, and in 'critical theory' generally,[3] as is exemplified by Marcuse's (1964) study of the ideology of advanced industrial society, and by some of the writings of Habermas.[4] A further impetus has been given to Marxist studies of culture, and in particular of literature and the mass media, by the structuralist movement. The major achievement in this field to date is undoubtedly the work of Goldmann, from his study of Pascal, Racine, and Jansenism, to his later essays on cultural creation and the sociology of the novel (see Goldmann, 1964, 1977, 1981 and the text below). The concentration of attention upon cultural analysis and ideology-critique in much recent Marxist theory has itself attracted criticism from those who see in this orientation a distorted view of the ways in which class domination is maintained, in so far as this is explained by the effects of a dominant ideology rather than by economic control and political power.[5]

But although there have been substantial Marxist contributions to the sociology of literature, and more broadly to aesthetic theory, as well as some new analyses of ideology in general, major deficiencies remain in Marxist cultural studies as a whole. In the field of law no work has appeared which could take its place beside Renner's pioneering study; the study of religion has been largely neglected, and no Marxist work is at all comparable with the researches of Durkheim and Weber;[6] and surprisingly little attention has been given to the analysis of political ideologies, such as nationalism or individualism. Outside the sphere of literature the principal Marxist contribution has been to the sociology of science, where two distinct lines of inquiry can be traced; one concerned with the social context of science,[7] the other with science and technology as

[3] On this, see Held (1980).

[4] In recent works, however, he has emphasized more strongly the economic and political dimensions of 'late capitalism'; see Habermas (1976) and the text in Part VII below.

[5] See Abercrombie, Hill, and Turner (1980). It should be noted, however, that it is not only Marxist cultural studies which have flourished in recent decades; there has also been a notable revival of Marxist political economy.

[6] But in recent Marxist anthropology, influenced by Lévi-Strauss' studies of myth, some interesting studies have begun; see especially, Godelier (1977), pt. iii.

[7] See Bernal (1939), Rose and Rose (1976).

ideology.[8] It is no doubt the profound social consequences of the rapid advance of science (not least in the creation of increasingly destructive weapons), as well as the traditionally high valuation of science in Marxist thought, which account for the attention that has been given to this field. But as we have shown, there are other important regions of culture to which Marxist sociology, as it develops further, should turn its attention.

[8] See Habermas (1970) and (1972).

KARL MARX
*
Material Activity, Consciousness, and Ideology*

The fact is, therefore, that determinate individuals, who are productively active in a definite way, enter into these determinate social and political relations. Empirical observation must, in each particular case, show empirically, and without any mystification or speculation, the connection of the social and political structure with production. The social structure and the State are continually evolving out of the life-process of determinate individuals, of individuals not as they may appear in their own or other people's imagination, but as they really are: i.e. as they act, produce their material life, and are occupied within determinate material limits, presuppositions, and conditions, which are independent of their will.

The production of ideas, conceptions, and consciousness is at first directly interwoven with the material activity and the material intercourse of men, the language of real life. Representation and thought, the mental intercourse of men, still appear at this stage as the direct emanation of their material behaviour. The same applies to mental production as it is expressed in the political, legal, moral, religious, and metaphysical language of a people. Men are the producers of their conceptions, ideas, etc.,—real, active men, as they are conditioned by a determinate development of their productive forces, and of the intercourse which corresponds to these, up to its most extensive forms. Consciousness can never be anything else than conscious existence, and the existence of men is their actual life-process. If in all ideology men and their circumstances appear upside down as in a *camera obscura*, this phenomenon arises from their historical life process just as the inversion of objects on the retina does from their physical life-process.

In direct contrast to German philosophy, which descends from

---

* From Karl Marx and Friedrich Engels, *The German Ideology* (1846), vol. i, pt. IA. Translated by Tom Bottomore.

heaven to earth, here we ascend from earth to heaven. That is to say, we do not set out from what men say, imagine, or conceive, nor from what has been said, thought, imagined, or conceived of men, in order to arrive at men in the flesh. We begin with real, active men, and from their real life-process show the development of the ideological reflexes and echoes of this life-process. The phantoms of the human brain also are necessary sublimates of men's material life-process, which can be empirically established and which is bound to material preconditions. Morality, religion, metaphysics, and other ideologies, and their corresponding forms of consciousness, no longer retain therefore their appearance of autonomous existence. They have no history, no development; it is men, who, in developing their material production and their material intercourse, change, along with this their real existence, their thinking and the products of their thinking. Life is not determined by consciousness, but consciousness by life. Those who adopt the first method of approach begin with consciousness, regarded as the living individual; those who adopt the second, which corresponds with real life, begin with the real living individuals themselves, and consider consciousness only as *their* consciousness.

This method of approach is not without presuppositions, but it begins with the real presuppositions and does not abandon them for a moment. Its premises are men, not in some imaginary condition of fulfilment or stability, but in their actual, empirically observable process of development under determinate conditions. As soon as this active life-process is delineated, history ceases to be a collection of dead facts as it is with the empiricists (themselves still abstract), or an illusory activity of illusory subjects, as with the idealists.

Where speculation ends—in real life—real, positive science, the representation of the practical activity and the practical process of development of men, begins. Phrase-making about consciousness ceases, and real knowledge has to take its place. When reality is depicted, philosophy as an independent activity loses its medium of existence. At the most its place can only be taken by a conspectus of the general results, which are derived from the consideration of the historical development of men. In themselves and detached from real history, these abstractions have not the least value. They can only serve to facilitate the arrangement of historical material, and to indicate the sequence of its separate layers. They do not in the least provide, as does philosophy, a recipé or schema, according to which the epochs of history can rightly be distinguished. On the contrary, the difficulties only begin when we set about the consideration and arrangement of the material, whether of a past epoch or of the present, and the representation of reality.

KARL MARX
*
The Arts and the Development of Society*

In the case of the arts it is well known that some 'golden ages' are quite unrelated to the general development of society, and hence also to its material basis, the skeletal structure of its organization. For example, the Greeks compared with the moderns, or with Shakespeare. It is recognized that certain forms of art, e.g. the epic, can no longer be produced in their world epoch-making and classical shape once the production of art as such begins; consequently, that within the sphere of the arts themselves certain important forms are only possible at an immature stage of artistic development. If this is the case with the relation between different kinds of art in the artistic sphere itself, then it is already less remarkable that it should be so in the relation of the realm of art as a whole to the general development of society. The difficulty consists only in getting a general grasp of these contradictions. As soon as they have been specified they are already accounted for.

Let us take, for example, the relation of Greek art, and then that of Shakespeare, to the present time. It is well known that Greek mythology is not only the arsenal of Greek art, but also its foundation. Is the view of nature and of social relations which underlie Greek imagination, and therefore Greek [mythology], possible with self-acting mule spindles and railways and locomotives and the electric telegraph? Where is Vulcan in face of Roberts and Co., Jupiter against the lightning conductor, or Hermes against the Crédit Mobilier? All mythology overcomes,

* From Karl Marx, *Grundrisse der Kritik der politischen Ökonomie* (1953 edn.), pp. 30–1. Translated by Tom Bottomore.

masters and shapes the forces of nature in the imagination and through the imagination; it vanishes therefore with real mastery over them. What becomes of Fama alongside Printing House Square? Greek art presupposes Greek mythology, i.e. nature and the social forms already reworked in an unconsciously artistic way by the popular imagination. That is its material. Not any mythology one pleases; i.e. not some arbitrarily chosen, unconsciously artistic reworking of nature (which here comprises everything objective, hence including society). Egyptian mythology could never have been the soil or the womb of Greek art. But in any case, *a* mythology. Hence, in no way a development of society which excludes any mythological, mythologizing relation to nature, or demands of the artist an imagination that is independent of mythology.

From another side: is Achilles possible with powder and shot? Or the *Iliad* in general with the printing press and printing machines? Does not song and saga and the muse necessarily come to an end with the composing stick, and do not the necessary conditions of epic poetry therefore necessarily disappear?

But the difficulty does not lie in understanding that Greek art and the epic are bound up with specific forms of social development. The difficulty is that they still afford us artistic pleasure and in certain respects stand as a norm and an unattainable model.

A man cannot become a child again, or he becomes childish. But does not the *naïveté* of the child give him pleasure, and must he himself not strive to reproduce its truth at a higher level? Is not the true character of every epoch brought to life again in the nature of its children? Why should not the historical childhood of humanity, where it unfolded in its most beautiful form, exercise an eternal charm as a stage which will never return? There are ill-mannered children and precocious children. Many ancient peoples fall into this category. The Greeks were normal children. The charm of their art for us is not in contradiction with the undeveloped stage of society from which it grew. It is its result, rather, and inextricably bound up with the fact that the immature social conditions from which it emerged, and could only emerge, can never return.

ANTONIO GRAMSCI

*

The Intellectuals*

Are intellectuals an autonomous and independent social group, or does every social group have its own particular specialized category of intellectuals? The problem is a complex one, because of the variety of forms assumed to date by the real historical process of formation of the different categories of intellectuals.

The most important of these forms are two:

1. Every social group, coming into existence on the original terrain of an essential function in the world of economic production, creates together with itself, organically, one or more strata[1] of intellectuals which give it homogeneity and an awareness of its own function not only in the economic but also in the social and political fields. The capitalist entrepreneur creates alongside himself the industrial technician, the specialist in political economy, the organizers of a new culture, of a new legal system, etc. It should be noted that the entrepreneur himself represents a higher level of social elaboration, already characterized by a certain directive [*dirigente*] and technical (i.e. intellectual) capacity: he must have a certain technical capacity, not

---

* From Antonio Gramsci, *Selections from the Prison Notebooks*, edited and translated by Quintin Hoare and Geoffrey Nowell Smith (London: Lawrence and Wishart, 1971), pp. 5–14. Reprinted by permission of Lawrence and Wishart.

---

[1] The Italian word here is '*ceti*' which does not carry quite the same connotations as 'strata', but which we have been forced to translate in that way for lack of alternatives. It should be noted that Gramsci tends, for reasons of censorship, to avoid using the word class in contexts where its Marxist overtones would be apparent, preferring (as for example in this sentence) the more neutral 'social group'. The word 'group', however, is not always a euphemism for 'class', and to avoid ambiguity Gramsci uses the phrase 'fundamental social group' when he wishes to emphasize the fact that he is referring to one or other of the major social classes (bourgeoisie, proletariat) defined in strict Marxist terms by its position in the fundamental relations of production. Class groupings which do not have this fundamental role are often described as 'castes' (aristocracy, etc.). The word 'category', on the other hand, which also occurs on this page, Gramsci tends to use in the standard Italian sense of members of a trade or profession, though also more generally. Throughout this edition we have rendered Gramsci's usage as literally as possible. (Translators' note).

Culture and Ideology 169

only in the limited sphere of his activity and initiative but in other spheres as well, at least in those which are closest to economic production. He must be an organizer of masses of men; he must be an organiser of the 'confidence' of investors in his business, of the customers for his product, etc.

If not all entrepreneurs, at least an élite amongst them must have the capacity to be an organizer of society in general, including all its complex organism of services, right up to the state organism, because of the need to create the conditions most favourable to the expansion of their own class; or at the least they must possess the capacity to choose the deputies (specialized employees) to whom to entrust this activity of organizing the general system of relationships external to the business itself. It can be observed that the 'organic' intellectuals which every new class creates alongside itself and elaborates in the course of its development, are for the most part 'specializations' of partial aspects of the primitive activity of the new social type which the new class has brought into prominence.[2]

Even feudal lords were possessors of a particular technical capacity, military capacity, and it is precisely from the moment at which the aristocracy loses its monopoly of technico-military capacity that the crisis of feudalism begins. But the formation of intellectuals in the feudal world and in the preceding classical world is a question to be examined separately: this formation and elaboration follows ways and means which must be studied concretely. Thus it is to be noted that the mass of the peasantry, although it performs an essential function in the world of

---

[2] Mosca's *Elementi di Scienza Politica* (new expanded edition, 1923) are worth looking at in this connection. Mosca's so-called 'political class' is nothing other than the intellectual category of the dominant social group. Mosca's concept of 'political class' can be connected with Pareto's concept of the *élite*, which is another attempt to interpret the historical phenomenon of the intellectuals and their function in the life of the state and of society. Mosca's book is an enormous hotch-potch, of a sociological and positivistic character, plus the tendentiousness of immediate politics which makes it less indigestible and livelier from a literary point of view. (Gramsci's note.) The term 'political class' is usually translated in English as 'ruling class', which is also the title of the English version of Mosca's *Elementi* (G. Mosca, *The Ruling Class*, New York 1939). Gaetano Mosca (1858–1941) was, together with Pareto and Michels, one of the major early Italian exponents of the theory of political *élites*. Although sympathetic to fascism, Mosca was basically a conservative, who saw the *élite* in rather more static terms than did some of his fellows. (Translators' note.)

production, does not elaborate its own 'organic' intellectuals, nor does it 'assimilate' any stratum of 'traditional' intellectuals, although it is from the peasantry that other social groups draw many of their intellectuals and a high proportion of traditional intellectuals are of peasant origin.[3]

2. However, every 'essential' social group which emerges into history out of the preceding economic structure, and as an expression of a development of this structure, has found (at least in all of history up to the present) categories of intellectuals already in existence and which seemed indeed to represent an historical continuity uninterrupted even by the most complicated and radical changes in political and social forms.

The most typical of these categories of intellectuals is that of the ecclesiastics, who for a long time (for a whole phase of history, which is partly characterized by this very monopoly) held a monopoly of a number of important services: religious ideology, that is the philosophy and science of the age, together with schools, education, morality, justice, charity, good works, etc. The category of ecclesiastics can be considered the category of intellectuals organically bound to the landed aristocracy. It had equal status juridically with the aristocracy, with which it shared the exercise of feudal ownership of land, and the use of state privileges connected with property.[4] But the monopoly held by

[3] Notably in Southern Italy. See below, 'The Different Position of Urban and Rural-type Intellectuals', pp. 14–23. Gramsci's general argument, here as elsewhere in the *Quaderni*, is that the person of peasant origin who becomes an 'intellectual' (priest, lawyer, etc.) generally thereby ceases to be organically linked to his class of origin. One of the essential differences between, say, the Catholic Church and the revolutionary party of the working class lies in the fact that, ideally, the proletariat should be able to generate its own 'organic' intellectuals within the class and who remain intellectuals *of* their class. (Translators' note.)

[4] For one category of these intellectuals, possibly the most important after the ecclesiastical for its prestige and the social function it performed in primitive societies, the category of *medical men* in the wide sense, that is all those who 'struggle' or seem to struggle against death and disease, compare the *Storia della medicina* of Arturo Castiglioni. Note that there has been a connection between religion and medicine, and in certain areas there still is: hospitals in the hands of religious orders for certain organizational functions, apart from the fact that wherever the doctor appears, so does the priest (exorcism, various forms of assistance, etc.). Many great religious figures were and are conceived of as great 'healers': the idea of miracles, up to the resurrection of the dead. Even in the case of kings the belief long survived that they could heal with the laying on of hands, etc. (Gramsci's note.)

the ecclesiastics in the superstructural field[5] was not exercised without a struggle or without limitations, and hence there took place the birth, in various forms (to be gone into and studied concretely), of other categories, favoured and enabled to expand by the growing strength of the central power of the monarch, right up to absolutism. Thus we find the formation of the *noblesse de robe*, with its own privileges, a stratum of administrators, etc., scholars and scientists, theorists, non-ecclesiastical philosophers, etc.

Since these various categories of traditional intellectuals experience through an *esprit de corps* their uninterrupted historical continuity and their special qualification, they thus put themselves forward as autonomous and independent of the dominant social group. The self-assessment is not without consequences in the ideological and political field, consequences of wide-ranging import. The whole of idealist philosophy can easily be connected with this position assumed by the social complex of intellectuals and can be defined as the expression of that social utopia by which the intellectuals think of themselves as 'independent', autonomous, endowed with a character of their own, etc.

One should note however that if the Pope and the leading hierarchy of the Church consider themselves more linked to Christ and to the apostles than they are to senators Agnelli and Benni,[6] the same does not hold for Gentile and Croce, for example: Croce in particular feels himself closely linked to Aristotle and Plato, but he does not conceal, on the other hand, his links with senators Agnelli and Benni, and it is precisely here that one can discern the most significant character of Croce's philosophy.

What are the 'maximum' limits of acceptance of the term 'intellectual'? Can one find a unitary criterion to characterize equally all the diverse and disparate activities of intellectuals and to distinguish these at the same time and in an essential way from

---

[5] From this has come the general sense of 'intellectual' or 'specialist' of the word *chierico* (clerk, cleric) in many languages of romance origin or heavily influenced, through church Latin, by the romance languages, together with its correlative *laico* (lay, layman) in the sense of profane, non-specialist. (Gramsci's note.)

[6] Heads of FIAT and Montecatini (Chemicals) respectively. For Agnelli, of whom Gramsci had direct experience during the *Ordine Nuovo* period, see note 11 on p. 286. (Translators' note.)

the activities of other social groupings? The most widespread error of method seems to me that of having looked for this criterion of distinction in the intrinsic nature of intellectual activities, rather than in the ensemble of the system of relations in which these activities (and therefore the intellectual groups who personify them) have their place within the general complex of social relations. Indeed the worker or proletarian, for example, is not specifically characterized by his manual or instrumental work, but by performing this work in specific conditions and in specific social relations (apart from the consideration that purely physical labour does not exist and that even Taylor's phrase of 'trained gorilla'[7] is a metaphor to indicate a limit in a certain direction: in any physical work, even the most degraded and mechanical, there exists a minimum of technical qualification, that is, a minimum of creative intellectual activity.) And we have already observed that the entrepreneur, by virtue of his very function, must have to some degree a certain number of qualifications of an intellectual nature although his part in society is determined not by these, but by the general social relations which specifically characterize the position of the entrepreneur within industry.

All men are intellectuals, one could therefore say: but not all men have in society the function of intellectuals.[8]

When one distinguishes between intellectuals and non-intellectuals, one is referring in reality only to the immediate social function of the professional category of the intellectuals, that is, one has in mind the direction in which their specific professional activity is weighted, whether towards intellectual elaboration or towards muscular-nervous effort. This means that, although one can speak of intellectuals, one cannot speak of non-intellectuals, because non-intellectuals do not exist. But even the relationship between efforts of intellectual-cerebral elaboration and muscular-nervous effort is not always the same, so that there are varying degrees of specific intellectual activity. There is no human activity from which every form of intellectual partici-

---

[7] For Frederick Taylor and his notion of the manual worker as a 'trained gorilla', see Gramsci's essay *Americanism and Fordism*, pp. 277–318 of this volume. (Translators' note.)

[8] Thus, because it can happen that everyone at some time fries a couple of eggs or sews up a tear in a jacket, we do not necessarily say that everyone is a cook or a tailor. (Gramsci's note.)

pation can be excluded: *Homo faber* cannot be separated from *Homo sapiens*.[9] Each man, finally, outside his professional activity, carries on some form of intellectual activity, that is, he is a 'philosopher', an artist, a man of taste, he participates in a particular conception of the world, has a conscious line of moral conduct, and therefore contributes to sustain a conception of the world or to modify it, that is, to bring into being new modes of thought.

The problem of creating a new stratum of intellectuals consists therefore in the critical elaboration of the intellectual activity that exists in everyone at a certain degree of development, modifying its relationship with the muscular-nervous effort towards a new equilibrium, and ensuring that the muscular-nervous effort itself, in so far as it is an element of a general practical activity, which is perpetually innovating the physical and social world, becomes the foundation of a new and integral conception of the world. The traditional and vulgarized type of the intellectual is given by the man of letters, the philosopher, the artist. Therefore journalists, who claim to be men of letters, philosophers, artists, also regard themselves as the 'true' intellectuals. In the modern world, technical education, closely bound to industrial labour even at the most primitive and unqualified level, must form the basis of the new type of intellectual.

On this basis the weekly *Ordine Nuovo*[10] worked to develop certain forms of new intellectualism and to determine its new concepts, and this was not the least of the reasons for its success, since such a conception corresponded to latent aspirations and conformed to the development of the real forms of life. The mode of being of the new intellectual can no longer consist in eloquence, which is an exterior and momentary mover of feelings and passions, but in active participation in practical life, as constructor, organizer, 'permanent persuader' and not just a simple orator (but superior at the same time to the abstract mathematical spirit); from technique-as-work one proceeds to technique-as-science and to the humanistic conception of history,

---

[9] i.e. Man the maker (or tool-bearer) and Man the thinker. (Translators' note.)

[10] The *Ordine Nuovo*, the magazine edited by Gramsci during his days as a militant in Turin, ran as a 'weekly review of Socialist culture' in 1919 and 1920. (Translators' note.)

without which one remains 'specialized' and does not become 'directive'[11] (specialized and political).

Thus there are historically formed specialized categories for the exercise of the intellectual function. They are formed in connection with all social groups, but especially in connection with the more important, and they undergo more extensive and complex elaboration in connection with the dominant social group. One of the most important characteristics of any group that is developing towards dominance is its struggle to assimilate and to conquer 'ideologically' the traditional intellectuals, but this assimilation and conquest is made quicker and more efficacious the more the group in question succeeds in simultaneously elaborating its own organic intellectuals.

The enormous development of activity and organization of education in the broad sense in the societies that emerged from the medieval world is an index of the importance assumed in the modern world by intellectual functions and categories. Parallel with the attempt to deepen and to broaden the 'intellectuality' of each individual, there has also been an attempt to multiply and narrow the various specializations. This can be seen from educational institutions at all levels, up to and including the organisms that exist to promote so-called 'high culture' in all fields of science and technology.

School is the instrument through which intellectuals of various levels are elaborated. The complexity of the intellectual function in different states can be measured objectively by the number and gradation of specialized schools: the more extensive the 'area' covered by education and the more numerous the 'vertical' 'levels' of schooling, the more complex is the cultural world, the civilization, of a particular state. A point of comparison can be found in the sphere of industrial technology: the industrialization of a country can be measured by how well equipped it is in the production of machines with which to produce machines, and in the manufacture of ever more accurate instruments for making both machines and further instruments for making machines, etc.

---

[11] *Dirigente*. This extremely condensed and elliptical sentence contains a number of key Gramscian ideas: on the possibility of proletarian cultural hegemony through domination of the work process, on the distinction between organic intellectuals of the working class and traditional intellectuals from outside, on the unity of theory and practice as a basic Marxist postulate, etc. (Translators' note.)

The country which is best equipped in the construction of instruments for experimental scientific laboratories and in the construction of instruments with which to test the first instruments, can be regarded as the most complex in the technical-industrial field, with the highest level of civilization, etc. The same applies to the preparation of intellectuals and to the schools dedicated to this preparation; schools and institutes of high culture can be assimilated to each other. In this field also, quantity cannot be separated from quality. To the most refined technical-cultural specialization there cannot but correspond the maximum possible diffusion of primary education and the maximum care taken to expand the middle grades numerically as much as possible. Naturally this need to provide the widest base possible for the selection and elaboration of the top intellectual qualifications—i.e. to give a democratic structure to high culture and top-level technology—is not without its disadvantages: it creates the possibility of vast crises of unemployment for the middle intellectual strata, and in all modern societies this actually takes place.

It is worth noting that the elaboration of intellectual strata in concrete reality does not take place on the terrain of abstract democracy but in accordance with very concrete traditional historical processes. Strata have grown up which traditionally 'produce' intellectuals and these strata coincide with those which have specialized in 'saving', i.e. the petty and middle landed bourgeoisie and certain strata of the petty and middle urban bourgeoisie. The varying distribution of different types of school (classical and professional)[12] over the 'economic' territory and the varying aspirations of different categories within these strata determine, or give form to, the production of various branches of intellectual specialization. Thus in Italy the rural bourgeoisie produces in particular state functionaries and professional people, whereas the urban bourgeoisie produces technicians for industry. Consequently it is largely northern Italy which produces technicians and the South which produces functionaries and professional men.

[12] The Italian school system above compulsory level is based on a division between academic ('classical' and 'scientific') education and vocational training for professional purposes. Technical and, at the academic level, 'scientific' colleges tend to be concentrated in the Northern industrial areas. (Translators' note.)

The relationship between the intellectuals and the world of production is not as direct as it is with the fundamental social groups but is, in varying degrees, 'mediated' by the whole fabric of society and by the complex of superstructures, of which the intellectuals are, precisely, the 'functionaries'. It should be possible both to measure the 'organic quality' [*organicità*] of the various intellectual strata and their degree of connection with a fundamental social group, and to establish a gradation of their functions and of the superstructures from the bottom to the top (from the structural base upwards). What we can do, for the moment, is to fix two major superstructural 'levels': the one that can be called 'civil society', that is the ensemble of organisms commonly called 'private', and that of 'political society' or 'the State'. These two levels correspond on the one hand to the function of 'hegemony' which the dominant group exercises throughout society and on the other hand to that of 'direct domination' or command exercised through the State and 'juridical' government. The functions in question are precisely organizational and connective. The intellectuals are the dominant group's 'deputies' exercising the subaltern functions of social hegemony and political government. These comprise:

1. The 'spontaneous' consent given by the great masses of the population to the general direction imposed on social life by the dominant fundamental group; this consent is 'historically' caused by the prestige (and consequent confidence) which the dominant group enjoys because of its position and function in the world of production.

2. The apparatus of state coercive power which 'legally' enforces discipline on those groups who do not 'consent' either actively or passively. This apparatus is, however, constituted for the whole of society in anticipation of moments of crisis of command and direction when spontaneous consent has failed.

This way of posing the problem has as a result a considerable extension of the concept of intellectual, but it is the only way which enables one to reach a concrete approximation of reality. It also clashes with preconceptions of caste. The function of organizing social hegemony and state domination certainly gives rise to a particular division of labour and therefore to a whole hierarchy of qualifications in some of which there is no apparent attribution of directive or organizational functions. For example, in the apparatus of social and state direction there exist a whole series of jobs of a manual and instrumental character (non-executive work, agents rather than officials or functionaries).[13] It is obvious that such a distinction has to be made just as it is obvious that other distinctions have to be made as well. Indeed, intellectual activity must also be distinguished in terms of its intrinsic characteristics, according to levels which in moments of extreme opposition represent a real qualitative difference—at the highest level would be the creators of the various sciences, philosophy, art, etc., at the lowest the most humble 'administrators' and divulgators of pre-existing, traditional, accumulated intellectual wealth.[14]

In the modern world the category of intellectuals, understood in this sense, has undergone an unprecedented expansion. The democratic-bureaucratic system has given rise to a great mass of functions which are not all justified by the social necessities of production, though they are justified by the political necessities of the dominant fundamental group. Hence Loria's[15] conception of the unproductive 'worker' (but unproductive in relation to whom and to what mode of production?), a conception which could in part be justified if one takes account of the fact that these masses exploit their position to take for themselves a large cut out of the national income. Mass formation has standardized individuals both psychologically and in terms of individual qualification and has produced the same phenomena as with other standardized masses: competition which makes necessary organizations for the defence of professions, unemployment, over-production in the schools, emigration, etc.

[13] *funzionari*: in Italian usage the word is applied to the middle and higher echelons of the bureaucracy. Conversely 'administrators' (*'amministratori'*) is used here (end of paragraph) to mean people who merely 'administer' the decisions of others. The phrase 'non-executive work' is a translation of [*impiego*] *di ordine e non di concetto* which refers to distinctions within clerical work. (Translators' note.)

[14] Here again military organization offers a model of complex gradations between subaltern officers, senior officers and general staff, not to mention the NCO's, whose importance is greater than is generally admitted. It is worth observing that all these parts feel a solidarity and indeed that it is the lower strata that display the most blatant *esprit de corps*, from which they derive a certain 'conceit' which is apt to lay them open to jokes and witticisms. (Gramsci's note.) 'Conceit' (*boria*) is a reference to an idea of Vico. (Translators' note.)

[15] The notion of the 'unproductive labourer' is not in fact an invention of Loria's but has its origins in Marx's definitions of productive and unproductive labour in *Capital*, which Loria, in his characteristic way both vulgarized and claimed as his own discovery. (Translators' note.)

LUCIEN GOLDMANN

Problems of a Sociology of the Novel*

But the first problem that a sociology of the novel should have confronted is that of the relation between the *novel form* itself and the *structure* of the social environment in which it developed, that is to say, between the novel as a literary genre and individualistic modern society.

It seems to me today that a combination of the analyses of Lukács and Girard,[1] even though they were both developed without specifically sociological preoccupations, makes it possible, if not to elucidate this problem entirely, at least to make a decisive step towards its elucidation.

I have just said that the novel can be characterized as the story of a search for authentic values in a degraded mode, in a degraded society, and that this degradation, in so far as it concerns the hero, is expressed principally through the mediatization, the reduction of authentic values to the implicit level, and their disappearance as manifest realities. This is obviously a particularly complex structure and it would be difficult to imagine that it could one day emerge simply from individual invention without any basis in the social life of the group.

What, however, would be quite inconceivable, is that a literary form of such dialectical complexity should be rediscovered, over a period of centuries, among the most different writers in the most varied countries, that it should have become the form *par excellence* in which was expressed, on the literary plane, the content of a whole period, without there being either a homology or a significant relation between this form and the most important aspects of social life.

This hypothesis seems to me particularly simple and above all productive and credible, though it has taken me years to find it.

The novel form seems to me, in effect, to be *the transposition on the*

---

* From Lucien Goldmann, *Towards a Sociology of the Novel* (1964). English trans. by Alan Sheridan (London: Tavistock Publications, 1975), pp. 6–17. Copyright © Editions Gallimard, 1964. Translation © Tavistock Publications Ltd., 1975. Reprinted by permission of Associated Book Publishers Ltd.

---

[1] The reference is to Lukács (1916) and René Girard (1961) [Eds.].

literary plane of everyday life in the individualistic society created by market production. There is a *rigorous homology* between the literary form of the novel, as I have defined it with the help of Lukács and Girard, and the everyday relation between man and commodities in general, and by extension between men and other men, in a market society.

The natural, healthy relation between men and commodities is that in which production is consciously governed by future consumption, by the concrete qualities of objects, by their *use value*.

Now what characterizes market production is, on the contrary, the elimination of this relation with men's consciousness, its reduction to the implicit through the mediation of the new economic reality created by this form of production: *exchange value* (. . .)

Of course, use values continue to exist and even to govern, in the last resort, the whole of the economic life; but their action assumes an *implicit character, exactly like that of authentic values in the fictional world*.

On the conscious, manifest plane, *the economic life* is composed of people orientated exclusively towards exchange values, degraded values, to which are added in production a number of individuals—the creators in every sphere—who remain essentially orientated towards use values and who by virtue of that fact are situated on the fringes of society and become *problematic individuals*; and, of course, even these individuals unless they accept the romantic illusion (Girard would say lie) of the *total* rupture between essence and appearance, between the inner life and the social life, cannot be deluded as to the degradations that their creative activity undergoes in a market society, when this activity is manifested externally, when it becomes a book, a painting, teaching, a musical composition, etc., enjoying a certain prestige, and having therefore a certain price. It should be added that as the ultimate consumer, opposed in the very act of exchange to the producers, any individual in a market society finds himself at certain moments of the day aiming at qualitative use values that he can obtain only through the mediation of exchange values.

In view of this, there is nothing surprising about the creation of the novel as a literary genre. Its apparently extremely complex form is the one in which men live every day, when they are obliged to seek all quality, all use value in a mode degraded by the mediation of quantity, of exchange value—and this in a society in which any effort to orientate oneself *directly* towards use value can only produce individuals who are themselves degraded, but in a different mode, that of *the problematic individual*.

Thus the two structures, that of an important fictional genre and that of exchange proved to be strictly homologous, to the point at which one might speak of one and the same structure manifesting itself on two different planes. Furthermore, as we shall see later, the *evolution* of the fictional form that corresponds to the world of reification can be understood only in so far as it is related to a *homologous history* of the structure of reification.

However, before making a few remarks about this homology between the two evolutions we must examine the problem, particularly important for the sociologist, of the process by which the literary form was able to emerge out of the economic reality, and of the modifications that the study of this process forces us to introduce into the traditional representation of the sociological conditioning of literary creation.

One fact is striking at the outset; the traditional scheme of literary sociology, whether Marxist or not, cannot be applied in the case of the structural homology just referred to. Most work in the sociology of literature established a relation between the most important literary works and the collective *consciousness* of the particular social group from which they emerged. On this point, the traditional Marxist position does not differ essentially from non-Marxist sociological work as a whole, in relation to which it introduces only four new ideas, namely:

(a) The literary work is not the mere reflection of a real, given collective consciousness, but the culmination at a very advanced level of coherence of tendencies peculiar to the consciousness of a particular group, a consciousness that must be conceived as a dynamic reality, orientated towards a certain state of equilibrium. What really separates, in this as in all other spheres, Marxist sociology from positivistic, relativist, or eclectic sociological tendencies is the fact that it sees the key concept not in the *real* collective consciousness, but in the constructed concept (*zugerechnet*) of *possible consciousness* which, alone, makes an understanding of the first possible.

(b) The relation between collective ideology and great individual literary, philosophical, theological etc. creations resides not in an identity of content, but in a more advanced coherence and in a homology of structures, which can be expressed in imaginary contents very different from the real content of the collective consciousness.

(c) The work corresponding to the mental structure of the particular social group may be elaborated in certain exceptional cases by an individual with very few relations with this group. The *social* character of the work resides above all in the fact that an individual can never establish by himself a coherent mental structure corresponding to what is called a 'world view'. Such a structure can be elaborated only by a group, the individual being capable only of carrying it to a very high degree of coherence and transposing it on the level of imaginary creation, conceptual thought, etc.

(d) The collective consciousness is neither a primary reality, nor an autonomous reality; it is elaborated implicitly in the overall behaviour of individuals participating in the economic, social, political life, etc.

These are evidently extremely important theses, sufficient to establish a very great difference between Marxist thinking and other conceptions of the sociology of literature. Nevertheless, despite these differences, Marxist theoreticians, like positivistic or relativistic sociologists of literature, have always thought that the social life can be expressed on the literary, artistic, or philosophical plane only through the intermediary link of the collective consciousness.

In the case we have just studied, however, what strikes one first is the fact that although we find a strict homology between the structures of economic life and a certain particularly important manifestation, one can detect no analogous structure at the level of the *collective consciousness* that seemed hitherto to be the indispensable intermediary link to realize either the homology or an intelligible, significant relation between the different aspects of social existence.

The novel analysed by Lukács and Girard no longer seems to be the imaginary transposition of the *conscious structures* of a particular group, but seems to express on the contrary (and this may be the case of a very large part of modern art in general) a search for values that no social group defends effectively and that the economic life tends to make implicit in all members of the society.

The old Marxist thesis whereby the proletariat was seen as the only social group capable of constituting the basis of a new culture, by virtue of the fact that it was not integrated into the reified society, set out from the traditional sociological representation that presupposed that all authentic, important cultural creation could emerge only from a fundamental harmony between the mental structure of the creator and that of a partial group of relative size, but universal ambition. In reality, for Western society at least, the Marxist analysis has proved inadequate; the Western proletariat, far from remaining alien to the reified society and opposing it as a revolutionary force, has on the contrary become integrated into it to a large degree, and its trade union and political action, far from overthrowing this society and replacing it by a socialist world, has enabled it to gain a relatively better place in it than Marx's analysis foresaw.

Furthermore, cultural creation, although increasingly threatened by the reified society, has continued to flourish. Fictional literature, as perhaps modern poetic creation and contemporary painting, are authentic forms of cultural creation even though they cannot be atttached to the consciousness—even a potential one—of a particular social group.

Before embarking on a study of the processes that made possible and produced this *direct* transposition of the economic life into the literary life, we should perhaps remark that although such a process seems contrary to the whole tradition of Marxist studies of cultural creation, it confirms nevertheless, in a quite unexpected way, one of the most important Marxist analyses of bourgeois thought, namely the theory of the fetishization of merchandise and reification. This analysis, which Marx regarded as one of his most important discoveries, affirms in effect that in market societies (that is to say, in types of society in which economic activity predominates), the collective consciousness gradually loses all active reality and tends to become a mere reflection[2] of the economic life and, ultimately, to disappear.

[2] I speak of a 'consciousness-reflection' when the content of this consciousness and the set of relations between the different elements of the content (what I call its structure) undergo the action of certain other domains of the social life, without acting in turn on them. In practice, this situation has probably never been reached in capitalist society. This society creates, however, a tendency to the rapid and gradual diminution of the action of consciousness on the economic life and, conversely, to a continual increase of the action of the economic sector of the social life on the content and structure of consciousness.

Culture and Ideology 183

There was obviously, therefore, between this *particular* analysis of Marx and the general theory of literary and philosophical creation of later Marxists, who presupposed an active role of the collective consciousness, not a contradiction but an incoherence. The latter theory never envisaged the consequences for the sociology of literature of Marx's belief that there survives in market societies a radical modification of the status of the individual and collective consciousness and, implicitly, relations between the infrastructure and the superstructure. The analysis of reification elaborated first by Marx on the level of everyday life, then developed by Lukács in the field of philosophical, scientific, and political thought, finally taken up by a number of theoreticians in various specific domains, and about which I have myself published a study, would appear therefore, for the moment at least, to be confirmed by the facts in the sociological analysis of a certain fictional form.

Having said this, the question arises as to how the link between the economic structures and literary manifestations is made in a society in which this link occurs *outside the collective consciousness*.

With regard to this I have formulated the hypothesis of the convergent action of four different factors, namely:

(a) The birth in the thinking of members of bourgeois society, on the basis of economic behaviour and the existence of exchange value, of the *category of mediation* as a fundamental and increasingly developed form of thought, with an implicit tendency to replace this thought by a total false consciousness in which the mediating value becomes an absolute value and in which the mediated value disappears entirely or, to put it more clearly, with the tendency to conceive of the access to all values from the point of view of mediation, together with a propensity to make of money and social prestige absolute values and not merely mediations that provide access to other values of a qualitative character.

(b) The survival in this society of a number of individuals who are essentially *problematic* in so far as their thinking and behaviour remain dominated by qualitative values, even though they are unable to extract themselves entirely from the existence of the degrading mediation whose action permeates the whole of the social structure.

These individuals include, above all, the creators, writers, artists, philosophers, theologians, men of action, etc., whose thought and behaviour are governed above all by the quality of their work even though they cannot escape entirely from the action of the market and from the welcome extended them by the reified society.

(c) Since no important work can be the expression of a purely individual experience, it is likely that the novel genre could emerge and be developed only in so far as a *non-conceptualized*, affective discontent, an affective aspiration towards qualitative values, was developed either in society as a whole, or perhaps solely among the middle strata from which most novelists have come.³

(d) Lastly, in the liberal market societies, there was a set of values, which, though not trans-individual, nevertheless, had a universal aim and, within these societies, a general validity. These were the values of liberal individualism that were bound up with the very existence of the competitive market (in France, liberty, equality, and property, in Germany, *Bildungsideal*, with their derivatives, tolerance, the rights of man, development of the personality, etc.). On the basis of these values, there developed the category of *individual* biography that became the constitutive element of the novel. Here, however, it assumed the form of the *problematic* individual, on the basis of the following:

1. the personal experience of the problematic individuals mentioned above under (b);

---

³ There arises a problem here that is difficult to solve at the moment, but which might one day be solved by concrete sociological research. I mean the problem of the collective, affective, non-conceptualized 'sound-box' that made possible the development of the novel form.

Initially, I thought that reification, while tending to dissolve and to integrate in the over-all society different partial groups, and, therefore, to deprive them to a certain extent of their specificity, had a character so contrary to both the biological and psychological reality of the individual human being that it could not fail to engender in *all* individual human beings, to a greater or lesser degree, reactions of opposition (or, if this reification becomes degraded in a qualitatively more advanced way, to reactions of evasion), thus creating a diffuse resistance to the reified world, a resistance that would constitute the background of fictional creation.

Later, however, it seemed to me that this hypothesis contained an unproved *a priori* supposition: that of the existence of a biological nature whose external manifestations could not be entirely denatured by social reality.

In fact, it is just as likely that resistances, even affective ones, to reification are circumscribed within certain particular social strata, which positive research ought to delimit.

2. the internal contradiction between individualism as a universal value produced by bourgeois society and the important and painful limitations that this society itself brought to the possibilities of the development of the individual.

This hypothetical schema seems to me to be confirmed among other things by the fact that, when one of these four elements, individualism, has gradually been eliminated by the transformation of the economic life and the replacement of the economy of free competition by an economy of cartels and monopolies (a transformation that began at the end of the nineteenth century, but whose qualitative turning-point most economists would place between 1900 and 1910), we witness a parallel transformation of the novel form that culminates in the gradual dissolution and disappearance of the individual character, of the hero; a transformation that seems to me to be characterized in an extremely schematic way by the existence of two periods:

(a) The first, transitional period, during which the disappearance of the importance of the individual brings with it attempts to replace biography as the content of the work of fiction with values produced by different ideologies. For although, in Western societies, these values have proved to be too weak to produce their own literary forms, they might well give a new lease of life to an already existing form that was losing its former content. First and foremost, on this level, are the ideas of community and collective reality (institutions, family, social group, revolution, etc.) that had been introduced and developed in Western thinking by the socialist ideology.

(b) The second period, which begins more or less with Kafka and continues to the contemporary *nouveau roman*, and which has not yet come to an end, is characterized by an abandonment of any attempt to replace the problematic hero and individual biography by another reality and by the effort to write the novel of the absence of the subject, of the non-existence of any ongoing search.[4]

---

[4] Lukács characterized the time of the traditional novel by the proposition: 'We have started on our way, our journey is over.' One might characterize the new novel by the suppression of the first half of this statement. Its time might be characterized by the statement: 'The aspiration is there, but the journey is over' (Kafka, Nathalie Sarraute), or simply by the observation that 'the journey is already over, though we never started on our way' (Robbe-Grillet's first three novels).

It goes without saying that this attempt to safeguard the novel form by giving it a content, related no doubt to the content of the traditional novel (it had always been the literary form of the problematic search and the absence of positive values), but nevertheless essentially different (it now involves the elimination of two essential elements of the specific content of the novel: the psychology of the problematic hero and the story of his demoniacal search), was to produce at the same time parallel orientations towards different forms of expression. There may be here elements for a sociology of the theatre of absence (Beckett, Ionesco, Adamov during a certain period) and also of certain aspects of non-figurative painting.

Lastly, we should mention a problem that might and ought to be the subject of later research. The novel form that we have just studied is essentially critical and oppositional. It is a form of resistance to developing bourgeois society. An individual resistance that can fall back, without a group, only on *affective* and *non-conceptualized* psychical processes precisely because conscious resistances that might have elaborated literary forms implying the possibility of a positive hero (in the first place, a proletarian oppositional consciousness such as Marx had hoped for and predicted) had not become sufficiently developed in Western societies. The novel with a problematic hero thus proves, contrary to traditional opinion, to be a literary form bound up certainly with history and the development of the bourgeoisie, but not the expression of the real or possible consciousness of that class.

But the problem remains as to whether, parallel with this literary form, there did not develop other forms that might correspond to the conscious values and effective aspirations of the bourgeoisie; and, on this point, I should like to mention, merely as a general and hypothetical suggestion, the possibility that the work of Balzac—whose structure ought, indeed, to be analysed from this point of view—might constitute the only great literary expression of the world as structured by the conscious values of the bourgeoisie: individualism, the thirst for power, money, and eroticism, which triumph over the ancient feudal values of altruism, charity, and love.

Sociologically, this hypothesis, if it proves to be correct, might be related to the fact that the work of Balzac is situated precisely at a period in which individualism, ahistorical in itself, structured the consciousness of a bourgeoisie that was in the process of constructing a new society and found itself at the highest and most intense level of its real historical efficacity.

We should also ask ourselves why, with the exception of this single case, this form of fictional literature had only a secondary importance in the history of Western culture, why the real consciousness and aspirations of the bourgeoisie never succeeded again, in the course of the nineteenth and twentieth centuries, in creating a literary form of its own that might be situated on the same level as the other forms that constitute the Western literary tradition.

On this point, I would like to make a few general hypotheses. The analysis that I have just developed extends to one of the most important novel forms a statement that now seems to me to be valid for almost all forms of *authentic cultural creation*. In relation to this statement the only expression that I could see for the moment was constituted precisely by the work of Balzac,[5] who was able to create a great literary universe structured by purely individualistic values, at a historical moment when, concurrently, men animated by ahistorical values were accomplishing a considerable historical upheaval (an upheaval that was not really completed in France until the end of the bourgeois revolution in 1848). With this single exception (but perhaps one should add a few other possible exceptions that may have escaped my attention), it seems to me that there is valid literary and artistic creation only when there is an aspiration to transcendence on the part of the individual and a search for qualitative trans-individual values. 'Man passes beyond man', I have written, slightly altering Pascal. This means that man can be authentic only in so far as he conceives himself or feels himself as part of a developing whole and situates himself in a historical or transcendent trans-individual dimension. But bourgeois ideology, bound up like bourgeois society itself with the existence of

[5] A year ago, when dealing with the same problems and mentioning the existence of the novel with a problematic hero and of a fictional sub-literature with a positive hero, I wrote, 'Lastly, I shall conclude this article with a great question mark, that of the sociological study of the works of Balzac. These works, it seems to me, constitute a novel form of their own, one that integrates important elements belonging to the two types of novels that I have mentioned and probably represents the most important form of fictional expression in history.'

The remarks formulated in these pages are an attempt to develop in greater detail the hypothesis hinted at in these lines.

economic activity, is precisely the first ideology in history that is both radically profane and ahistorical; the first ideology whose tendency is to deny anything sacred, whether the otherworldly sacredness of the transcendent religions or the immanent sacredness of the historical future. It is, it seems to me, the fundamental reason why bourgeois society created the first radically non-aesthetic form of consciousness. The essential character of bourgeois ideology, rationalism, ignores in its extreme expressions the very existence of art. There is no Cartesian or Spinozian aesthetics, or even an aesthetics for Baumgarten—art is merely an inferior form of knowledge.

It is no accident therefore if, with the exception of a few particular situations, we do not find any great literary manifestations of the bourgeois consciousness itself. In a society bound up with the market, the artist is, as I have already said, a problematic individual, and this means a critical individual, opposed to society.

Nevertheless reified bourgeois ideology had its thematic values, values that were sometimes authentic, such as those of individualism, sometimes purely conventional, which Lukács called false consciousness and, in their extreme forms, bad faith, and Heidegger's 'chatter'. These stereotypes, whether authentic or conventional, thematized in the collective consciousness, were later able to produce, side by side with the authentic novel form, a parallel literature that also recounted an individual history and, naturally enough, since conceptualized values were involved, could depict a positive hero.

WALTER BENJAMIN

*

*The Work of Art in the Age of Mechanical Reproduction**

Preface

When Marx undertook his critique of the capitalistic mode of production, this mode was in its infancy. Marx directed his efforts

* From Walter Benjamin, *Illuminations*. Translated by Harry Zohn edited by Hannah Arendt (New York: Harcourt Brace Jovanovich, 1968; London: Cape,

in such a way as to give them prognostic value. He went back to the basic conditions underlying capitalistic production and through his presentation showed what could be expected of capitalism in the future. The result was that one could expect it not only to exploit the proletariat with increasing intensity, but ultimately to create conditions which would make it possible to abolish capitalism itself.

The transformation of the superstructure, which takes place far more slowly than that of the substructure, has taken more than half a century to manifest in all areas of culture the change in the conditions of production. Only today can it be indicated what form this has taken. Certain prognostic requirements should be met by these statements. However, theses about the art of the proletariat after its assumption of power or about the art of a classless society would have less bearing on these demands than theses about the developmental tendencies of art under present conditions of production. Their dialectic is no less noticeable in the superstructure than in the economy. It would therefore be wrong to underestimate the value of such theses as a weapon. They brush aside a number of outmoded concepts, such as creativity and genius, eternal value and mystery—concepts whose uncontrolled (and at present almost uncontrollable) application would lead to a processing of data in the Fascist sense. The concepts which are introduced into the theory of art in what follows differ from the more familiar terms in that they are completely useless for the purposes of Fascism. They are, on the other hand, useful for the formulation of revolutionary demands in the politics of art.

I

In principle a work of art has always been reproducible. Manmade artifacts could always be imitated by men. Replicas were made by pupils in practice of their craft, by masters for diffusing their works, and, finally, by third parties in the pursuit of gain. Mechanical reproduction of a work of art, however, represents something new. Historically, it advanced intermittently and in leaps at long intervals, but with accelerated intensity. The Greeks knew only two procedures of technically

1970). Copyright © 1955 by Suhrkamp Verlag, Frankfurt A.M.; English translation copyright © 1968 by Harcourt Brace Jovanovich, Inc. Reprinted by permission of Harcourt Brace Jovanovich, Inc., and Jonathan Cape Ltd.

reproducing works of art: founding and stamping. Bronzes, terra cottas, and coins were the only art works which they could produce in quantity. All others were unique and could not be mechanically reproduced. With the woodcut graphic art became mechanically reproducible for the first time, long before script became reproducible by print. The enormous changes which printing, the mechanical reproduction of writing, has brought about in literature are a familiar story. However, within the phenomenon which we are here examining from the perspective of world history, print is merely a special, though particularly important, case. During the Middle Ages engraving and etching were added to the woodcut; at the beginning of the nineteenth century lithography made its appearance.

With lithography the technique of reproduction reached an essentially new stage. This much more direct process was distinguished by the tracing of the design on a stone rather than its incision on a block of wood or its etching on a copperplate and permitted graphic art for the first time to put its products on the market, not only in large numbers as hitherto, but also in daily changing forms. Lithography enabled graphic art to illustrate everyday life, and it began to keep pace with printing. But only a few decades after its invention, lithography was surpassed by photography. For the first time in the process of pictorial reproduction, photography freed the hand of the most important artistic functions which henceforth devolved only upon the eye looking into a lens. Since the eye perceives more swiftly than the hand can draw, the process of pictorial reproduction was accelerated so enormously that it could keep pace with speech. A film operator shooting a scene in the studio captures the images at the speed of an actor's speech. Just as lithography virtually implied the illustrated newspaper, so did photography foreshadow the sound film. The technical reproduction of sound was tackled at the end of the last century. These convergent endeavours made predictable a situation which Paul Valéry pointed up in this sentence: 'Just as water, gas, and electricity are brought into our houses from far off to satisfy our needs in response to a minimal effort, so we shall be supplied with visual or auditory images, which will appear and disappear at a simple movement of the hand, hardly more than a sign' (*op. cit.*, p. 226).[1]

---

[1] Valéry, 'La conquête de l'ubiquité', in (1931) [Eds.].

Around 1900 technical reproduction had reached a standard that not only permitted it to reproduce all transmitted works of art and thus to cause the most profound change in their impact upon the public; it also had captured a place of its own among the artistic processes. For the study of this standard nothing is more revealing than the nature of the repercussions that these two different manifestations—the reproduction of works of art and the art of the film—have had on art in its traditional form. (. . .)

XII

Mechanical reproduction of art changes the reaction of the masses toward art. The reactionary attitude toward a Picasso painting changes into the progressive reaction toward a Chaplin movie. The progressive reaction is characterized by the direct, intimate fusion of visual and emotional enjoyment with the orientation of the expert. Such fusion is of great social significance. The greater the decrease in the social significance of an art form, the sharper the distinction between criticism and enjoyment by the public. The conventional is uncritically enjoyed, and the truly new is criticized with aversion. With regard to the screen, the critical and the receptive attitudes of the public coincide. The decisive reason for this is that individual reactions are predetermined by the mass audience response they are about to produce, and this is nowhere more pronounced than in the film. The moment these responses become manifest they control each other. Again, the comparison with painting is fruitful. A painting has always had an excellent chance to be viewed by one person or by a few. The simultaneous contemplation of paintings by a large public, such as developed in the nineteenth century, is an early symptom of the crisis of painting, a crisis which was by no means occasioned exclusively by photography but rather in a relatively independent manner by the appeal of art works to the masses.

Painting simply is in no position to present an object for simultaneous collective experience, as it was possible for architecture at all times, for the epic poem in the past, and for the movie today. Although this circumstance in itself should not lead one to conclusions about the social role of painting, it does constitute a serious threat as soon as painting, under special conditions and, as it were, against its nature, is confronted directly by the masses. In the churches and monasteries of the Middle Ages and at the princely courts up to the end of the eighteenth century, a collective reception of paintings did not occur simultaneously, but by graduated and hierarchized mediation. The change that has come about is an expression of the particular conflict in which painting was implicated by the mechanical reproducibility of paintings. Although paintings began to be publicly exhibited in galleries and salons, there was no way for the masses to organize and control themselves in their reception. Thus the same public which responds in a progressive manner toward a grotesque film is bound to respond in a reactionary manner to surrealism. (...)

Epilogue

The growing proletarianization of modern man and the increasing formation of masses are two aspects of the same process. Fascism attempts to organize the newly created proletarian masses without affecting the property structure which the masses strive to eliminate. Fascism sees its salvation in giving these masses not their right, but instead a chance to express themselves.[2] The masses have a right to change property relations; Fascism seeks to give them an expression while preserving property. The logical result of Fascism is the introduction of aesthetics into political life. The violation of the masses, whom Fascism, with its *Führer* cult, forces to their knees, has its counterpart in the violation of an apparatus which is pressed into the production of ritual values.

All efforts to render politics aesthetic culminate in one thing: war. War and war only can set a goal for mass movements on the largest scale while respecting the traditional property system.

---

[2] One technical feature is significant here, especially with regard to newsreels, the propagandist importance of which can hardly be overestimated. Mass reproduction is aided especially by the reproduction of masses. In big parades and monster rallies, in sports events, and in war, all of which nowadays are captured by camera and sound recording, the masses are brought face to face with themselves. This process, whose significance need not be stressed, is intimately connected with the development of the techniques of reproduction and photography. Mass movements are usually discerned more clearly by a camera than by the naked eye. A bird's-eye view best captures gatherings of hundreds of thousands. And even though such a view may be as accessible to the human eye as it is to the camera, the image received by the eye cannot be enlarged the way a negative is enlarged. This means that mass movements, including war, constitute a form of human behaviour which particularly favours mechanical equipment.

This is the political formula for the situation. The technological formula may be stated as follows: Only war makes it possible to mobilize all of today's technical resources while maintaining the property system. It goes without saying that the Fascist apotheosis of war does not employ such arguments. Still, Marinetti says in his manifesto on the Ethiopian colonial war:

For twenty-seven years we Futurists have rebelled against the branding of war as antiaesthetic. . . . Accordingly we state: . . . War is beautiful because it establishes man's dominion over the subjugated machinery by means of gas masks, terrifying megaphones, flamethrowers, and small tanks. War is beautiful because it initiates the dreamt-of metalization of the human body. War is beautiful because it enriches a flowering meadow with the fiery orchids of machine guns. War is beautiful because it combines the gunfire, the cannonades, the cease-fire, the scents, and the stench of putrefaction into a symphony. War is beautiful because it creates new architecture, like that of the big tanks, the geometrical formation flights, the smoke spirals from burning villages, and many others. . . . Poets and artists of Futurism! . . . remember these principles of an aesthetics of war so that your struggle for a new literature and a new graphic art . . . may be illumined by them!

This manifesto has the virtue of clarity. Its formulations deserve to be accepted by dialecticians. To the latter, the aesthetics of today's war appears as follows: If the natural utilization of productive forces is impeded by the property system, the increase in technical devices, in speed, and in the sources of energy will press for an unnatural utilization, and this is found in war. The destructiveness of war furnishes proof that society has not been mature enough to incorporate technology as its organ, that technology has not been sufficiently developed to cope with the elemental forces of society. The horrible features of imperialistic warfare are attributable to the discrepancy between the tremendous means of production and their inadequate utilization in the process of production—in other words, to unemployment and the lack of markets. Imperialistic war is a rebellion of technology which collects, in the form of 'human material', the claims to which society has denied its natural material. Instead of draining rivers, society directs a human stream into a bed of trenches; instead of dropping seeds from airplanes, it drops incendiary bombs over cities; and through gas warfare the aura is abolished in a new way.

Fiat ars—pereat mundus, says Fascism, and, as Marinetti admits, expects war to supply the artistic gratification of a sense perception that has been changed by technology. This is evidently the consummation of *l'art pour l'art*. Mankind, which in Homer's time was an object of contemplation for the Olympian gods, now is one for itself. Its self-alienation has reached such a degree that it can experience its own destruction as an aesthetic pleasure of the first order. This is the situation of politics which Fascism is rendering aesthetic. Communism responds by politicizing art.

OTTO BAUER

*

The Nation as a Cultural Community*

Capitalism has uprooted the rural population, torn people from the soil to which they had been bound ever since the earliest settlements, drawn them beyond the narrow confines of the parish. Their sons have moved into the town where people from distant parts of the country come together, influence one another, and mingle their blood; where the traditional way of life of the peasantry, which repeated itself endlessly through the changing seasons, is replaced by the vigorously pulsating life of the city—a new ever-changing world which destroys all their inherited attitudes. With the changes in the industrial situation they are flung now here, now there. What a difference there is between the modern metal-worker who works today for the great iron magnates on the Rhine and tomorrow is shunted off to Silesia by a wave of industrial change, who courts his wife in Saxony and brings up his children in Berlin, and his grandfather who spent his whole life from cradle to grave in a remote Alpine village, visited the neighbouring small town perhaps only twice a year on the occasion of an annual market or one of the important church festivals, and never got to know the peasants in the next village because a mountain range made communication between the villages difficult. How different too is our metal-worker's

* From Otto Bauer, *Die Nationalitätenfrage und die Sozialdemokratie* (Vienna: Wiener Volksbuchhandlung, 1907), pp. 87–94. Translated by Patrick Goode. By permission of Verlag der Wiener Volksbuchhandlung.

brother, who inherited his father's land in the mountain village. The old, traditional methods of running a farm have been subject to constant change and new initiatives under the influence of the agricultural co-operatives, itinerant instructors, agricultural shows, and the like; the peasant has become a businessman who knows how to calculate the price of his commodity, to bargain with the dealers from the towns, and to exploit the competition among them. He is a producer and seller of commodities just like the dealers and the manufacturers in the town, and now that he is linked with the town population by all the bonds of business transactions he is no longer beyond the reach of its cultural influence. Perhaps he already travels into town on a bicycle to bargain with his customers, and instead of the traditional costume he now wears urban clothes bought in the town, the style of which shows the influence, if not of the latest fashion from Paris or Vienna, at least of the one before that.

These psychological transformations produced by capitalist development, have affected our whole *educational system*, but they in turn would have been impossible without the development of the educational system. Education has become an essential instrument for modern development. A higher level of national education was required by modern capitalism because the complicated apparatus of administration of large scale enterprise was impossible without it; by the modern farmer because otherwise he could never have advanced to modern farming methods; by the modern state because without it local administration and a modern army could never have been created. Hence, the 19th century witnessed an impressive development of the national educational system. There is no need to elaborate the significance for the national cultural community of the fact that the working-class child in East Prussia and the peasant child in the Tyrol acquire from their reading books the same cultural elements, the same excerpts from our intellectual heritage, communicated in the same uniform German language.

What the educational system begins is continued by the *army*. The logical outcome of the system of conscription was compulsory military service. On the battlefields where the absolutist powers of old Europe were defeated by the French Revolution, the modern army was born; a people's army, not yet in terms of its purpose or its organization, but already in terms of its composition. Military service tears the peasant's son away from the narrow sphere of the village, brings him together with comrades from the town and from other parts of the country, and puts him under the influence of the population of the garrison town. Hence our military system, quite unintentionally, revolutionizes the mind. Not without reason is the man in Gerhart Hauptmann's *The Weavers*, who kindles to a fire the glowing sparks of rebellion, a soldier just returned home from the town.

The influence which school has on the child, military service on the young man, is consummated in the case of the adult by *democracy*. The freedom of association and assemblies, and freedom of the press, become a means for conveying the great issues of the day to every village and every workplace, for making world events influential in the destiny and the cultural outlook of every individual. Universal suffrage, which summons everyone to participate in decision making, obliges the political parties to fight for every last vote, while in the party slogans all the great achievements of our whole history and culture contend for the attention of every peasant and worker. Every speech, every newspaper, brings a part of our culture to each voter. And all of them, however different in origin, wealth, occupation, or political orientation, belong to a cultural community, because all of them, as objects of the struggle between parties, are subject to the same cultural influences, and within each individual this influence has taken effect and congealed into a character.

But of all the historical movements which together create the modern nation in the capitalist era by far the most important is the *labour movement*. Its direct influence is already extraordinarily great. It is this movement which has struggled to achieve for the workers at least such a reduction in hours of work that a part of our national cultural heritage can and does filter through to them too, and has raised their wages to the level where they are not completely excluded from the cultural life of the nation by sheer physical and spiritual pauperization. But this is not all. By arousing the fears of the property-owning classes threatened by socialism, it has forced them into battle; now the bourgeoisie too, and even the landowners, must try to influence the masses. They seek to organize workers for their own purposes, and to persuade artisans and peasants to unite in a struggle against the working class. Thus the conflict over the great question of property rages throughout society and around every individual. Through publications, associations, and newspapers, the arguments of the parties are brought to bear on every member of the public. Through party struggles some small part of the stream of our culture, however diluted, does reach everyone and affects their character; and this uniform cultural influence unites all of us in a self-contained cultural community.

At the time of Caesar the Germans were a cultural community, but this ancient community decayed when they became a settled people with the transition to agriculture. Local communities, sharply separated from each other in different places, from one valley to the next, replaced the national community. Only the ruling and property-owning classes were still united in a nation by a higher culture. *It was modern capitalism which recreated a genuine national culture of the whole people, transcending the narrow confines of the parish.* It achieved this by uprooting people, tearing them away from their local connections, and regrouping them geographically and occupationally in the process of forming modern classes and occupations. The task was accomplished by means of democracy (which is capitalism's own creation), schools, compulsory military service, and universal suffrage.

Is capitalism not entitled to be proud of its work? Though often traduced, has it not accomplished something tremendous in recreating the nation as a cultural community for all and not simply for the property-owning classes? No doubt. Yet capitalism should not set such a high value on its achievement. The rise of the modern national cultural community was made possible by *advances in the forces of production*. The steam engine which works for us, driving spinning machines and looms; the giant blast furnaces and Bessemer-converters busy on our behalf; the steamships and railways which have opened up the fruitful lands of the far corners of our earth to us; these have given the whole people that share in the products of culture which makes the nation a cultural community. It is the evolution of the forces of production, the machine, that we have to thank for that regrouping of the population from which our great wealth flows; the increased wealth has become a cultural capital which binds the people together in a cultural community. This development of the forces of production did indeed occur through capitalism; but the fact that it took place in this way also sets limits to the development of a national cultural community. The growth of our productive forces and thereby our wealth was a condition for the development of the modern nation; but the fact that so far it is only through capitalism in the service of capital, that these productive forces have been able to develop, sets limits to the participation of the masses in the culture of the nation and to the development of a national cultural community.

The development of the forces of production means a prodigious increase in the productivity of labour, but only a small part of the increased wealth goes to the masses who create it. Ownership of the means of labour has become an instrument for appropriating a major part of the constantly growing wealth. The worker spends only a part of his working day producing the goods which represent his own wealth; the rest of his day is spent creating that wealth which falls to the owner of the means of production. But material wealth is always transformed into cultural possessions. *The law of our present age* is thus that *the work of some becomes the culture of others.* (. . .)

But capitalism does not only inhibit the development of the whole people into a national cultural community directly through the fact of exploitation; it also does so indirectly by *the necessity of defending exploitation*. True, it developed *schools* so far as it needed them, but it will take care not to create a genuine system of national education which might give the masses full possession of the cultural heritage. Not only because it limits the period of schooling, so that it can exploit children and not waste their potential, nor simply because it resents the cost of education and would rather offer up its wealth for the instruments of its power, but above all because if the masses were to be educated to the point where they participated fully in the national culture they would not put up with capitalist domination for a day longer. Capitalism fears the schools because they educate its opponents, and so attempts to reduce them to means of maintaining its rule. Capitalism was obliged to introduce *compulsory military service*, but it has not created a people's army. It shuts its soldiers away in barracks to remove them as far as possible from the influence of the people, and by means of external insignia, geographical segregation, and indoctrination with its own ideology, it tries to create status sentiments which keep them at a distance from the life of the people. Capitalism was the creator of *democracy*. But democracy was the youthful love of the bourgeoisie; it is the fear of its old age because it has become the instrument through which the working class can achieve power.

Certainly we should rejoice at every attempt to communicate

to the workers some part of our science or art. But only visionaries will forget that although an exceptionally gifted individual worker can even today become a highly cultivated person, full possession of our cultural heritage must necessarily remain beyond the reach of the mass of the people. (. . .) But of course the day is now closer than ever when the mass of the population will be able to lay their hands on this great wealth and make the cultural heritage, produced by the work of the whole people, the possession of the whole people. That day will see the emergence of a truly national cultural community.

FURTHER READING
*

Baxandall, Lee and Morawski, Stefan (eds.), *Marx and Engels on Literature and Art* (1974)
Brym, Robert *Intellectuals and Politics* (1979)
Fischer, Ernst *The Necessity of Art* (1963)
Larrain, Jorge *The Concept of Ideology* (1979)
Marcuse, Herbert *One-Dimensional Man: Studies in the Ideology of Advanced Industrial Society* (1964)
Prawer, S. S. *Karl Marx and World Literature* (1976)

PART VI
*
The Development of Society

PRESENTATION

*

Marxist sociology, as we argued in our introduction, combines structural and historical analysis. What it provides is, first, a comprehensive interpretation of the historical development of human societies (conceived as a succession of distinct types of society); second, an analysis of the transitions from one type of society to another; and third, a framework for understanding the stages of development within a particular social formation, e.g. modern capitalism (see Part VII below).

There underlies the Marxist theory of social development a conception of progress. Marx himself, in the Preface of 1859,[1] referred to 'progressive epochs in the economic formation of society', and to the conditions in which 'new, higher relations of production' could emerge; and in his introduction to the *Grundrisse* he noted as a point to be taken up again that the 'concept of progress' should not be 'conceived in the usual abstract fashion'.[2] In broad terms, Marx's view seems to have been that the general progressiveness of society resulted from the growth of human productive powers—capitalist society, in particular, having demonstrated what colossal 'productive forces slumbered in the lap of social labour'[3]—on the basis of which a more complex and refined civilization could develop,[4] eventually reaching a stage at which human creative powers in all spheres, not simply in material production, would be able to manifest themselves freely. It is from this perspective that Hobsbawm (see the text below) characterizes the general content of history as *progress*. In similar vein an outstanding Marxist

---

[1] Karl Marx, *Contribution to a Critique of Political Economy* (1859). Preface. See also his study of these epochs in the *Grundrisse* (1857–9. English trans. Harmondsworth: Penguin Books, 1973), pp. 471–514.

[2] Karl Marx, *Grundrisse*, op. cit., p. 109.

[3] *Communist Manifesto*.

[4] At the same time Marx insisted that there might well be a disproportion between the growth of productive powers and the development of civilization; this is the point of his critical note on the concept of progress in the *Grundrisse*, op. cit., p. 109. Earlier, in the *Economic and Philosophical Manuscripts* (1844), he set out to show how in modern capitalist society 'the *devaluation* of the human world increases in direct relation with the *increase in value* of the world of things'.

archaeologist outlined an analogy between organic evolution and cultural progress, and went on to describe the economic revolutions which promoted the early advances in civilization.[5] From a different aspect Habermas, in his general reassessment of historical materialism (see the text below), emphasizes the importance of recognizing the two dimensions of progress which were, as we have seen, clearly distinguished by Marx himself; namely, the development of productive forces, and the development of social institutions which have to be judged in terms of the level of civilization they attain.

We must now consider a particular feature of Marx's conception of social development which differentiates it sharply from what he regarded as the 'abstract' theories of progress; namely, its view of development as a discontinuous process in which relatively abrupt transitions occur from one type of society to another, through the political struggles between classes culminating in social revolutions (see the text by Adler below). One particular transition—from feudalism to capitalism—has been far more intensively studied than any other, and there is an excellent collection of essays (Sweezy *et al.* 1976) which brings out some of the fundamental problems concerning the 'law of motion' of feudal societies. Hilton, in his introduction to the volume, notes that a central issue in the debate is that of the 'prime mover'; whether an external force (such as the injection of merchant capital accumulated elsewhere) was needed for the initial development of capitalism, and on the other side, what was the strength of various internal forces—changing class relations, the growth of towns, demographic changes. Other historical transitions—the origins of slave society or of feudalism, the emergence and development of the Asiatic type of society—have been less closely studied, and Marxist discussion of them has been less intense.[6]

The question of stages of development within a particular type of society has been discussed primarily with reference to capitalism (see Part VII below) where it poses problems concerning the class structure and political movements. At the same time, however, it has brought into the forefront of Marxist

---

[5] Childe (1956).

[6] But see, on the origins of slavery, Finley (1980), and on the Asiatic type of society, Krader (1975, the text in Part II above) and Turner (1978).

debate a new problem—that of development in the more recent sense of 'developing countries' of the Third World—which is now the focus of many studies of the 'world capitalist system', 'neo-imperialism', and 'post-colonial societies'. Some of the principal issues raised by this debate are discussed in the text by Cardoso and Faletto below.

Marx's theory, as Berlin (1963) observed, created 'a wholly new attitude to social and historical questions . . . so opening new avenues of human knowledge'; and in particular, it had a major influence in establishing the discipline of social history, or as we would prefer to call it, historical sociology. As Marxist sociology, in association with anthropology and history, establishes itself more strongly it will extend still further our knowledge of how societies have developed and, we may hope, how their development can be directed by human beings towards greater well-being.

KARL MARX

*

The Historical Tendency of Capitalist Accumulation*

What does the primitive accumulation of capital, i.e. its historical genesis, resolve itself into? In so far as it is not a direct transformation of slaves and serfs into wage-labourers, and therefore a mere change of form, it only means the expropriation of the immediate producers, i.e. the dissolution of private property based on the labour of its owner.

Private property, as the anithesis to social, collective property, exists only where the means of labour and external conditions of labour belong to private individuals. But according as these private individuals are labourers or not labourers, private property has a different character. The innumerable shades, that it at first sight presents, correspond to the intermediate stages lying between these two extremes. The private property of the labourer in his means of production is the foundation of petty industry, and petty industry is an essential condition for the development of social production and of the free individuality of the labourer himself. Of course, this petty mode of production exists also under slavery, serfdom, and other states of dependence. But it flourishes, it lets loose its whole energy, it attains its full classical form, only where the labourer is the private owner of the means of labour which he uses; the peasant of the land which he cultivates, the artisan of the tool which he handles as a virtuoso. This mode of production presupposes parcelling out of the soil, and of the other means of production. As it excludes the concentration of these means of production, so also it excludes co-operation, division of labour within each separate process of production, the control over, and the productive application of, the forces of Nature by society, and the free development of the social productive powers. It is only compatible with a primitive and limited society and system of production. To perpetuate it would be, as Pecqueur rightly says, 'to decree universal medioc-

* From Karl Marx, *Capital*, vol. i, ch. 24, Section 7. Translated by Tom Bottomore. See also Marx's texts in Parts I and II above.

rity'. At a certain stage of development it brings forth the material agencies for its own dissolution. From that moment new forces and new passions spring up in the bosom of society; but the old social organization fetters them and keeps them down. It must be annihilated; it is annihilated. Its annihilation, the transformation of the individualized and scattered means of production into socially concentrated ones, of the pigmy property of the many into the huge property of the few, the expropriation of the great mass of the people from the soil, from the means of subsistence, and from the means of labour, this fearful and painful expropriation of the mass of the people forms the prelude to the history of capital. It comprises a series of forcible measures, of which we have passed in review only those that have been epoch-making as methods of the primitive accumulation of capital. The expropriation of the immediate producers was accomplished with merciless vandalism, and under the stimulus of the most infamous, sordid, petty, and odious passions. Self-earned private property that is based, so to say, on the fusing together of the isolated, independent labouring individual with the conditions of his labour, is supplanted by capitalist private property, which rests on exploitation of the nominally free labour of others.

As soon as this process of transformation has sufficiently decomposed the old society from top to bottom, as soon as the labourers are turned into proletarians, and their means of labour into capital, as soon as the capitalist mode of production stands on its own feet, then the further socialization of labour and further transformation of the land and other means of production into socially exploited and, therefore, common means of production, as well as the further expropriation of private proprietors, takes a new form. That which is now to be expropriated is no longer the labourer working for himself, but the capitalist exploiting many labourers. This expropriation is accomplished by the action of the immanent laws of capitalist production itself, by the centralization of capital. One capitalist always kills many. Hand in hand with this centralization, this expropriation of many capitalists by few, develop, on an everextending scale, the co-operative form of the labour process, the conscious application of science, the planned exploitation of the earth, the transformation of the instruments of labour into instruments which can only be used in co-operative work, the economizing of all means of production by their use as the means of production of combined, socialized labour, the entanglement of all peoples in the net of the world-market, and with this, the international character of the capitalist system. Along with the constantly diminishing number of the magnates of capital, who usurp and monopolize all the advantages of this process of transformation, grows the mass of misery, oppression, slavery, degradation, and exploitation; but with this too grows the revolt of the working-class, a class always increasing in numbers, and disciplined, united, organized by the mechanism of the process of capitalist production itself. The monopoly of capital becomes a fetter upon the mode of production, which has sprung up and flourished along with, and under it. Centralization of the means of production and socialization of labour at last reach a point where they become incompatible with their capitalist integument. The integument is burst asunder. The knell of capitalist private property sounds. The expropriators are expropriated.

ERIC HOBSBAWM

*

The Idea of Progress in Marx's Thought*

The *Formen* seek to formulate the *content* of history in its most general form. This content is *progress*. Neither those who deny the existence of historical progress nor those who (often basing themselves on the writings of the immature Marx) see Marx's thought merely as an ethical demand for the liberation of man, will find any support here. For Marx progress is something objectively definable, and at the same time pointing to what is desirable. The strength of the Marxist belief in the triumph of the free development of all men, depends not on the strength of Marx's hope for it, but on the assumed correctness of the analysis that this is indeed where historical development eventually leads mankind.

* From the Introduction by Eric Hobsbawm to Karl Marx, *Pre-Capitalist Economic Formations* (London: Lawrence & Wishart, 1964), pp. 12–16. Reprinted by permission of Lawrence & Wishart.

The objective basis of Marx's humanism, but of course also, and simultaneously, of his theory of social and economic evolution, is his analysis of man as a social animal. Man—or rather men—perform *labour*, i.e. they create and reproduce their existence in daily practice, breathing, seeking food, shelter, love, etc. They do this by operating *in* nature, taking from nature (and eventually consciously changing nature) for this purpose. This interaction between man and nature is, and produces, social evolution. Taking from nature, or determining the use of some bit of nature (including one's own body), can be, and indeed is in common parlance, seen as appropriation, which is therefore originally merely an aspect of labour. It is expressed in the concept of *property* (which is not by any means the same thing as the historically special case of *private* property). In the beginning, says Marx, 'the relationship of the worker to the objective conditions of his labour is one of ownership; this is the natural unity of labour with its material (*sachliche*) prerequisites' (p. 67). Being a social animal man develops both co-operation and a *social division of labour* (i.e. specialization of functions), which is not only made possible by, but increases the further possibilities of, producing a *surplus* over and above what is needed to maintain the individual and the community of which he is a part. The existence of both the surplus and the social division of labour makes possible *exchange*. But initially both production and exchange have as their object merely *use*—i.e. the maintenance of the producer and his community. These are the main analytical bricks out of which the theory is built, and all are in fact expansions or corollaries of, the original concept of man as a social animal of a special kind.[1]

Progress of course is observable in the growing emancipation of man from nature and his growing control over nature. This emancipation—i.e. from the situation as given when primitive men go about their living, and from the original and spontaneous (or as Marx says *naturwüchsig*—'as grown up in nature') relations

---

[1] For Engels's explanation of the evolution of man from apes, and hence of the difference between man and the other primates, cf. his 1876 draft on 'The part of labour in the transformation of the ape into man' in the *Dialectics of Nature*, Werke, xx, 444–55. [An English translation is published as an Appendix to Engels's *The Origin of the Family, Private Property and the State*, edited, with an Introduction, by Eleanor Burke Leacock (New York: International Publishers, 1972), pp. 251–64. Eds.]

which emerge from the process of the evolution of animals into human groups—affects not only the forces but also the relations of production. And it is with the latter aspect that the *Formen* deals. On the one hand, the relations men enter into as a result of the specialization of labour—and notably *exchange*—are progressively clarified and sophisticated, until the invention of *money* and with it of *commodity production* and exchange, provides a basis for procedures unimaginable before, including capital accumulation. This process, while mentioned at the outset of the present essay (p. 67), is not its major subject. On the other, the double relation of labour-property is progressively broken up, as man moves further from the *naturwüchsig* or spontaneously evolved primitive relation with nature. It takes the form of a progressive 'separation of free labour from the objective conditions of its realization—from the means of labour (*Arbeitsmittel*) and the material of labour. . . . Hence, above all, the separation of the labourer from the earth as his natural laboratory' (p. 67). Its final clarification is achieved under capitalism, when the worker is reduced to nothing but labour-power, and conversely, we may add, property to a control of the means of production entirely divorced from labour, while in the process of production there is a total separation between use (which has no direct relevance) and exchange and accumulation (which is the direct object of production). This is the process which, in its possible variations of type, Marx attempts to analyse here. Though particular social-economic formations, expressing particular phases of this evolution, are very relevant, it is the entire process, spanning the centuries and continents, which he has in mind. Hence his framework is chronological only in the broadest sense, and problems of, let us say, the transition from one phase to another, are not his primary concern, except in so far as they throw light on the long-term transformation.

But at the same time this process of the emancipation of man from his original natural conditions of production is one of human *individualization*. 'Man is only individualised (*vereinzelt sich*) through the process of history. He appears originally as a generic being, a tribal being, a herd animal. . . . Exchange itself is a major agent of this individualization. It makes the herd animal superfluous and dissolves it' (p. 96). This automatically implies a transformation in the relations of the individual to what was originally the community in which he functioned. The former community has been transmuted, in the extreme case of capitalism, into the dehumanized social mechanism which, while it actually makes individualization possible, is outside and hostile to the individual. And yet this process is one of immense possibilities for humanity. As Marx observes in a passage full of hope and splendour (pp. 84–5):

The ancient conception, in which man always appears (in however narrowly national, religious or political a definition) as the aim of production, seems very much more exalted than the modern world, in which production is the aim of man and wealth the aim of production. In fact, however, when the narrow bourgeois form has been peeled away, what is wealth, if not the universality of needs, capacities, enjoyments, productive powers, etc., of individuals, produced in universal exchange? What, if not the full development of human control over the forces of nature—those of his own nature as well as those of so-called 'nature'? What, if not the absolute elaboration of his creative dispositions, without any preconditions other than antecedent historical evolution which makes the totality of this evolution—i.e. the evolution of all human powers as such, unmeasured by any *previously established* yardstick—an end in itself? What is this, if not a situation where man does not reproduce himself in any determined form, but produces his totality? Where he does not seek to remain something formed by the past, but is in the absolute movement of becoming? In bourgeois political economy—and in the epoch of production to which it corresponds—this complete elaboration of what lies within man appears as the total alienation, and the destruction of all fixed, one-sided purposes as the sacrifice of the end in itself to a wholly external compulsion.[2]

Even in this most dehumanized and apparently contradictory form, the humanist ideal of free individual development is nearer than it ever was in all previous phases of history. It only awaits the passage from what Marx calls, in a lapidary phrase, the prehistoric stage of human society—the age of class societies of which capitalism is the last—to the age when man is in control of his fate, the age of communism.

Marx's vision is thus a marvellously unifying force. His model of social and economic development is one which (unlike Hegel's) can be applied to history to produce fruitful and original results rather than tautology; but at the same time it can be presented as the unfolding of the logical possibilities latent in a

---

[2] See the full text of this passage on pp. 278–9 below [Eds.].

few elementary and almost axiomatic statements about the nature of man—a dialectical working out of the contradictions of labour/property, and the division of labour.[3] It is a model of facts, but, seen from a slightly different angle, the *same* model provides us with value-judgments.

JÜRGEN HABERMAS
*
A Reconstruction of Historical Materialism*

To begin I shall examine the concepts of *social labour* and *history of the species*, as well as three fundamental assumptions of historical materialism.

1. *Socially organized labour* is the specific way in which humans, in contradistinction to animals, reproduce their lives.

Man can be distinguished from the animal by consciousness, religion, or anything else you please. He begins to distinguish himself from the animal the moment he begins to *produce* his means of subsistence, a step required by his physical organization. By producing food, man indirectly produces his material life itself.[1]

---

[3] Marx—unlike Hegel—is not taken in by the possibility—and indeed, at certain stages of thought, the necessity—of an abstract and *a priori* presentation of his theory. Cf. the section—brilliant, profound, and exciting as almost everything Marx wrote in this crucial period of his thought—on The Method of political economy, in the (unpublished) Introduction to the *Critique of Political Economy (Werke*, vol. xiii, 631–9), where he discusses the value of this procedure. [An English translation is published in Marx's *Grundrisse*, translated, with a Foreword, by Martin Nicolaus (Harmondsworth: Penguin Books, 1973), pp. 100–8. Eds.]

* From Jürgen Habermas, *Communication and the Evolution of Society*. Translated and with an Introduction by Thomas McCarthy (London: Heinemann, 1979), pp. 131–2, 138–42. German text copyright © 1976 by Suhrkamp Verlag. Introduction and English translation copyright © 1979 by Beacon Press. Reprinted by permission of Heinemann Educational Books and Beacon Press.

---

[1] Marx and Engels, *The German Ideology*, in Easton and Guddat (1967), p. 409.

The Development of Society 213

At a level of description that is unspecific in regard to the human mode of life, the exchange between the organism and its environment can be investigated in the physiological terms of material-exchange processes. But to grasp what is specific to the human mode of life, one must describe the relation between organism and environment at the level of labour processes. From the physical aspect the latter signify the expenditure of human energy and the transfer of energies in the economy of external nature; but what is decisive is the sociological aspect of the goal-directed transformation of material according to *rules of instrumental action*.[2]

Of course, under 'production' Marx understands not only the instrumental actions of a single individual, but also the *social cooperation* of different individuals:

The production of life, of one's own life in labor, and of another in procreation, now appears as a double relationship: on the one hand as a natural relationship, on the other as a social one. The latter is social in the sense that individuals co-operate, no matter under what conditions, in what manner, and for what purpose. Consequently a certain mode of production or industrial stage is always combined with a certain mode of co-operation or social stage, and this mode of co-operation is itself a 'productive force.' We observe in addition that the multitude of productive forces accessible to men determines the nature of society and that the 'history of mankind' must always be studied and treated in relation to the history of industry and exchange.[3]

The instrumental actions of different individuals are co-ordinated in a purposive-rational way, that is, with a view to the goal of production. The *rules of strategic action*, in accord with which cooperation comes about, are a necessary component of the labour process.

Means of subsistence are produced only to be consumed. The distribution of the product of labour is, like the labour itself, socially organized. In the case of rules of distribution, the concern is not with processing material or with the suitably coordinated application of means, but with the systematic connection of reciprocal expectations or interests. Thus the distribution of products requires rules of interaction that can be set intersubjectively at the level of linguistic understanding, detached

[2] On the delimitation of action types, cf. J. Habermas (1970), pp. 91 ff.
[3] Marx and Engels, *The German Ideology*, p. 421.

from the individual case, and made permanent as recognized norms or *rules of communicative action.*

We call a system that socially regulates labour and distribution an *economy*. According to Marx, then, the economic form of reproducing life is characteristic of the human stage of development. (. . .)

Marx links the concept of social labour with that of the *history of the species*. This phrase is intended in the first place to signal the materialist message that in the case of a single species natural evolution was continued by other means, namely, through the productive activity of the socialized individuals themselves. In sustaining their lives through social labour, men produce at the same time the material relations of life; they produce their society and the historical process in which individuals change along with their societies. The key to the reconstruction of the history of the species is provided by the concept of a *mode of production*. Marx conceives of history as a discrete series of modes of production, which, in its developmental-logical order, reveals the direction of social evolution. Let us recall the most important definitions.

A *mode of production* is characterized by a specific state of development of productive forces and by specific forms of social intercourse, that is, relations of production. The *forces of production* consist of (1) the labour power of those engaged in production, the producers; (2) technically useful knowledge in so far as it can be converted into instruments of labour that heighten productivity, that is, into technologies of production; (3) organizational knowledge in so far as it is applied to set labour power efficiently into motion, to qualify labour power, and to effectively co-ordinate the co-operation of labourers in accord with the division of labour (mobilization, qualification, and organization of labour power). Productive forces determine the degree of possible control over natural processes. On the other hand, the *relations of production* are those institutions and social mechanisms that determine the way in which (at a given stage of productive forces) labour power is combined with the available means of production. Regulation of access to the means of production, the way in which socially employed labour power is controlled, also determines indirectly the distribution of socially produced wealth. The relations of production express the distribution of social power; with the distributional pattern of socially recognized opportunities for need satisfaction, they prejudge the *interest structure* of a society. Historical materialism proceeds from the assumption that productive forces and productive relations do not vary independently, but form structures that (a) correspond with one another and (b) yield a finite number of structurally analogous stages of development, so that (c) there results a series of modes of production that are to be ordered in a developmental logic. (The handmill produces a society of feudal lords, the steam mill a society of industrial capitalists.)[4]

In the orthodox version, five modes of production are distinguished: (1) the primitive communal mode of bands and tribes prior to civilization; (2) the ancient mode based on slaveholding; (3) the feudal; (4) the capitalist; and finally (5) the socialist modes of production. The discussion of how the ancient Orient and the ancient Americas were to be ordered in this historical development led to the insertion of (6) an Asiatic mode of production.[5] These six modes of production are supposed to mark universal stages of social evolution. From an evolutionary stand-point, every particular *economic structure* can be analysed in terms of the various modes of production that have entered into a hierarchical combination in a historically concrete society. (A good example of this is Godelier's analysis of the Inca culture at the time of Spanish colonization.)[6]

The *dogmatic version* of the concept of a history of the species shares a number of weaknesses with eighteenth-century designs for a philosophy of history. The course of previous world history, which evidences a sequence of five or six modes of production, sets down the *unilinear, necessary, uninterrupted, and progressive development of a macrosubject*. I should like to oppose to this model of species history a weaker version, which is not open to the familiar criticisms of the objectivism of philosophy of history.[7]

(a) Historical materialism does not need to assume a *species-subject* that undergoes evolution. The bearers of evolution are

---

[4] Stalin (1974) *Dialectical and Historical Materialism*. This phrase about the hand mill and the steam mill comes from Marx, *The Poverty of Philosophy* (1847) [Eds.].

[5] J. Pecirka, 'Von der asiatischen Produktionsweise zu einer marxistischen Analyse der frühen Klassengesellschaften', *Eirene* 6 (Prague, 1967), pp. 141–74; and L. V. Danilova, 'Controversial Problems of the Theory of Precapitalist Societies', *Soviet Anthropology and Archaeology*, 9 (Spring 1971): 269–327.

[6] M. Godelier, *Perspectives in Marxist Anthropology* (1976).

[7] Recently, Marquardt, (1973).

rather societies and the acting subjects integrated into them; social evolution can be discerned in those structures that are replaced by more comprehensive structures in accord with a pattern that is to be rationally reconstructed. In the course of this structure-forming process, societies and individuals, together with their ego and group identities, undergo change. Even if social evolution should point in the direction of unified individuals consciously influencing the course of their own evolution, there would not arise any large-scale subjects, but at most self-established, higher-level, inter-subjective commonalities. (The specification of the concept of development is another question: in what sense can one conceive the rise of new structures as a movement?—only the empirical substrates are in motion.)[8]

(b) If we separate the logic from the dynamics of development—that is, the rationally reconstructible *pattern* of a hierarchy of more and more comprehensive structures from the *processes* through which the empirical substrates develop—then we need require of history neither unilinearity nor necessity, neither continuity nor irreversibility. We certainly do reckon with anthropologically deep-seated general structures, which were formed in the phase of hominization and which lay down the initial state of social evolution; these structures presumably arose to the extent that the cognitive and motivational potential of the anthropoid apes was transformed and reorganized under conditions of linguistic communication. These basic structures correspond, possibly, to the structures of consciousness that children today normally master between their fourth and seventh years, as soon as their cognitive, linguistic, and interactive abilities are integrated with one another.

Such structures describe the logical space in which more comprehensive structural formations can take shape; whether new structural formations arise at all, and if so, when, depends on *contingent* boundary conditions and on learning processes that can be investigated empirically. The genetic explanation of why a certain society has attained a certain level of development is independent of the structural explanation of how a system behaves—a system that conforms at every given stage to the logic of its acquired structures. Many paths can lead to the same level

---

[8] In an unpublished manuscript on the theory of evolution, Niklas Luhmann expresses doubts about the applicability of the concept of motion in this connection.

of development; *unilinear* developments are all the less probable, the more numerous the evolutionary units. Moreover, there is no guarantee of uninterrupted development; rather, it depends on accidental constellations whether a society remains unproductively stuck at the threshold of development or whether it solves its system problems by developing new structures. Finally, *retrogressions* in evolution are possible and in many cases empirically corroborated; of course, a society will not fall back behind a level of development, once it is established, without accompanying phenomena of forced regression; this can be seen, for example, in the case of Fascist Germany. It is not evolutionary processes that are *irreversible* but the structural sequences that a society must run through *if* and *to the extent that* it is involved in evolution.

3. Naturally the most controversial point is the *teleology* that, according to historical materialism, is supposed to be inherent in history. When we speak of evolution, we do in fact mean cumulative processes that exhibit a direction. Neo-evolutionism regards *increasing complexity* as an acceptable directional criterion. The more states a system can assume, the more complex the environment with which it can cope and against which it can maintain itself. Marx too attributed great significance to the category of the 'social division of labour'; by this he meant processes of system differentiation and of integration of functionally specified subsystems at a higher level, that is, processes that increase the internal complexity—and thereby the adaptive capacity—of a society. However, as a social-evolutionary directional criterion, complexity has a number of disadvantages:

(a) Complexity is a multidimensional concept. A society can be complex with respect to size, interdependence, and variability, with respect to achievements of generalization, integration, and respecification. As a result, complexity comparisons can become blurred, and questions of global classification from the viewpoint of complexity undecidable.[9]

(b) Moreover, there is no clear relation between complexity and self-maintenance. There are increases in complexity that turn out to be evolutionary dead ends. But without this connection, increases in complexity are unsuitable as directional signs; system complexity is equally ill-suited to be the basis for evolutionary stages of development.

(c) The connection between complexity and self-maintenance becomes problematic because societies, unlike organisms, do not have

[9] Luhmann points this out in the manuscript mentioned in n. 8.

clear-cut boundaries and objectively decidable problems of self-maintenance. The reproduction of societies is not measured in terms of rates of reproduction, that is, possibilities of the physical survival of their members, but in terms of securing a normatively prescribed societal identity, a culturally interpreted 'good' or 'tolerable' life.[10]

Marx judged social development not by increases in complexity but by the stage of development of productive forces and by the maturity of the forms of social intercourse.[11] The development of productive forces depends on the application of technically useful knowledge; and the basic institutions of a society embody moral-practical knowledge. Progress in these two dimensions is measured against the two universal validity claims we also use to measure the progress of empirical knowledge and of moral-practical insight, namely, the truth of propositions and the rightness of norms. I would like, therefore, to defend the thesis that the criteria of social progress singled out by historical materialism as the development of productive forces and the maturity of forms of social intercourse can be systematically justified.

MAX ADLER

*

Social Revolution*

But this only brings us to the threshold of the problem concerning the sociological meaning of revolution. So far we have only

[10] Cf. my critique of Luhmann in Habermas and Luhmann (1971), pp. 150 ff.; cf. (1973), pp. 66 ff.

[11] For example, H. Gericke, in 'Zur Dialektik von Produktivkraft und Produktionsverhältnis im Feudalismus'. *Zeitschrift für Geschichtswissenschaft*, 16 (1966): 914–32, distinguishes the 'increasingly higher degree of mastery of nature' from the 'increasingly maturer forms of corporate social life': 'The most important criteria and the decisive factors in historical progress are improvement of productive forces, especially the increase in conscious, goal-directed, success-oriented activity of immediate producers, as well as altered productive relations, which permit an ever increasing number of people to participate competently and actively in economic, social, political, and cultural processes' (pp. 918–19).

* From Max Adler, 'Zur Soziologie der Revolution' (1928). In *Austro-Marxism*, texts translated and edited by Tom Bottomore and Patrick Goode (Oxford: OUP, 1978), pp. 139–42. Copyright © Tom Bottomore and Patrick Goode, 1978. Reprinted by permission of Oxford University Press.

succeeded in showing that revolution *can* be a part of the necessary social process, not that it *must* be. At this point Marxist sociology makes an important conceptual distinction; between political and social revolutions. This distinction should not be regarded as political, although it has political consequences. It originates in a basic law of the social process; namely, its determination by economic development. In the famous sketch of the materialist conception of history which Marx gives in his Preface to the *Contribution to a Critique of Political Economy* we read:

At a certain stage of their development, the material forces of production come into conflict with the existing relations of production, or—what is merely a legal expression for the same thing—with the property relations within which they have previously moved. These relations change from forms of the development of the productive forces into fetters upon them. An epoch of social revolution then occurs.

This conception of social revolution is to be found also in the earlier writings of Marx, in fact if not in name; it occurs for the first time, notably, in the 'Critique of Hegel's Philosophy of Right', where it appears in the form of a distinction between human, and merely political, emancipation; and it underlies the whole train of thought of the *Communist Manifesto*. According to the definition given above, the social revolution is that social transformation which emerges from the insupportable contradiction between the forces of production and the relations of production. Hence it does not refer merely to a political reorganization of the state, but to the reconstruction of the foundations of a given social order. Compared with this, a political revolution is, as Marx himself put it, one which leaves the pillars of the house untouched. It does not change anything in the economic foundations of the social order, and attempts simply to modify the structure by redistributing power and thus changing the circle of those who are entitled to profit from society.

In order to understand the full significance of this distinction it is particularly necessary to keep in mind what was said previously, that it is a sociological, *not a political*, distinction. In this way, the whole concept of revolution is changed from the merely political idea of the transformation of the state, into the social concept of an economic change in the bases of society. The political changes which were previously considered pre-eminent are now seen as subordinate, and in spite of the political progress associated with them, only provisional. But if, in this way, every revolution points beyond its political effects to changes in society which it has either produced or prepared, then the distinction between political and social revolution must not be conceived as a means of historical classification, as though some revolutions were only political, and others only social. This cannot be the case, because every revolution is directed against an existing state order, and to that extent must be political. The distinction is intended rather to characterize certain active historical tendencies which appear to varying degrees in every revolution. In so far as a revolution includes efforts to change the social structure, then an essentially political revolution is also, to this extent, a social one. This is particularly true of the French Revolution which, by eliminating feudal and guild economic forms, broke the most burdensome fetters on the capitalist mode of production.

It follows that the concept of social revolution, in the sociological sense, is not simply a revolutionary demand addressed to the future, nor a teleological conception, nor merely fanatical idealism. It also applies to the past and has already been realized, at various levels, in the very diverse types of revolution that have occurred. In every case, the level depended upon the economic conditions in which the revolution took place. In particular, the varying levels of maturity of these conditions mean that a social revolution cannot attain its goal—the solution of basic economic contradictions—when this is not yet economically feasible. Marx's further significant distinction between a *total* and a merely *partial* revolution derives from this; the former being that which he calis a *radical* or *communist* revolution. This new definition indicates that the concept of social revolution is not identical with that of a Communist revolution; but the latter is the completion of the former.

But is not this concept of radical revolution, in the end just an idealistic, even Utopian, political programme, which is directed by a fanatical will, not by the requirements of the real historical process itself? This question can be answered by returning to Marx's definition of social revolution, according to which revolution is the consequence of a conflict in the economic structure of society. Such a conflict is only possible when the means of production cannot be adapted to the relations of production, because private ownership by a section of society precludes their planned utilization by the members of society as a whole. This means that the concept of social revolution presupposes a society in which there are economic class antagonisms. Like all social concepts, it is partly historical; social revolution is not a form of social development in general, but only that form through which the development of a class society must pass. Certainly, there will be revolutions in classless society too; but when the solidaristic character of this future society has become thoroughly established as a result of several generations growing up in the new conditions, and has undergone a comprehensive development, these revolutions will assume more the character of those intellectual revolutions that the scientific world, for example, experienced in the transition to the Copernican system, or through Einstein's theory. As the social revolution is born from class antagonism it becomes a particular form of the class struggle which permeates the whole history of class society. It is no longer an accidental, but a law-governed phenomenon; namely, the determined and necessary goal of the class struggle. In a class society, the class struggle is the only means of social progress since its object is to eliminate all class injustice and class domination. Hence social revolutions no longer appear merely as the outcome of the passions of specific periods of history, but as a measure of the conscious achievement of the historical progress that was economically possible at that time. But although revolutions can be seen, in this way, as inseparable elements in the continuous process of historical development, they do not thereby lose their character of a break with the present which can only be achieved by force; even if this might occur in the future through a simple democratic majority. In summary we can say: violent revolutions are necessary elements of the social process, so long as this proceeds through class antagonisms. Hence revolution itself undergoes development, in so far as it progresses from an essentially political to an essentially social form, and from a merely partial to a total transformation of society. In its application to the present level of economic development, in which the proletariat is the last economic class, this means that in the future a social revolution can only be a radical, that is, a Communist, revolution.

FERNANDO HENRIQUE CARDOSO and ENZO FALETTO

*

*Capitalism and Dependent Development**

In situations of extreme colonial dependency local history is almost reduced to a reflection of what happens in a metropolis. However, the decision by local forces to rebel against colonialism and to create a nation implies an attempt to influence local history according to local values and interests. Economic links with external markets still impose limits to decisions and actions even after independence. The contradiction between the attempt to cope with the market situation in a politically autonomous way and the *de facto* situation of dependency characterizes what is the specific ambiguity of nations where political sovereignty is expressed by the new state and where economic subordination is reinforced by the international division of labour and by the economic control exerted by former or new imperialist centres. From a sociological viewpoint, here is perhaps the core of the problem of national development in Latin America.

'National underdevelopment' is a situation of objective economic subordination to outside nations and enterprises and, at the same time, of partial political attempts to cope with 'national interests' through the state and social movements that try to preserve political autonomy. Ideological components play some role in the perception of what 'national interest' means, as well as in the rationalization about the possibility of the existence of nation-states that have submitted to foreign interests and pressures.

One of the aims of comprehensive analyses of the national development process is to determine the links between social groups that in their behaviour actually tie together the economic and political spheres. In so far as, by definition, links of economic dependency imply a relation between local and external classes, states, and enterprises, the analyses of local social and political

---

* From Fernando Henrique Cardoso and Enzo Faletto, *Dependency and Development in Latin America*. Translated by Marjory Mattingly Urquidi (Berkeley: University of California Press, 1979), pp. 21–4, 199–205. Copyright © 1979 by the Regents of the University of California. Reprinted by permission of the University of California Press.

groups must include the connections with international partners. Some local classes or groups sustain dependency ties, enforcing foreign economic and political interests. Others are opposed to the maintenance of a given pattern of dependency. Dependence thus finds not only internal 'expression', but also its true character as implying a situation that structurally entails a link with the outside in such a way that what happens 'internally' in a dependent country cannot be fully explained without taking into consideration the links that internal social groups have with external ones. Dependence should no longer be considered an 'external variable'; its analysis should be based on the relations between the different social classes within the dependent nations themselves.

This analysis does away with the idea that class relations in dependent countries are like those of the central countries during their early development. At the beginning of the development process in the central countries, market forces generally act as arbiter in the conflict of interests between the dominant groups. Thus, economic rationality, measured in money, was made a norm of society; and consumption and investment were limited by the growth of the economic system. Expansion of the system was due to a dynamic group that controlled investment decisions and imposed upon the entire society an orientation based on its own interests. The rising economic class possessed efficiency and consensus in capitalistic terms.

It was believed that the ruling groups expressed the general interest and that the market functioned adequately as a mechanism to satisfy general and particular interests. Other groups that exerted pressure in order to share in the fruits of 'progress' and in decision-making were ignored. Only long after the initial stage of industrialization did the popular classes participate politically and socially in the industrial societies.[1]

The national economies in the countries of 'early growth' succeeded in part because they were consolidated at the same time that the world market expanded, so that these countries came to occupy the leading positions in the system of international domination. From this scheme it is evident that 'early development', although a very broad and imprecise term, is significantly different from what has occurred in Latin America.

[1] See Alain Touraine, 'Industrialisation et conscience ouvrière a São Paulo', *Sociologie du Travail* (Apr. 1961).

It has been assumed that the peripheral countries would have to repeat the evolution of the economies of the central countries in order to achieve development. But it is clear that from its beginning the capitalist process implied an unequal relation between the central and the peripheral economies. Many 'underdeveloped' economies—as is the case of the Latin American—were incorporated into the capitalist system as colonies and later as national states, and they have stayed in the capitalist system throughout their history. They remain, however, peripheral economies with particular historical paths when compared with central capitalist economies.

Capitalism should be studied in the hope, not of finding how its history may repeat at a later date in the peripheral countries, but of learning how the relation between peripheral and central was produced. Although it is possible to distinguish in the economic history of Latin America the periods of mercantile, industrial, and financial capitalism, it is important for us to make clear what the relation of dependence meant in each of these phases. It would be senseless to seek how far or how close Latin American economies are from 'mercantilism', 'industrialism', or 'finance' forms of capitalism. They belong to the same international capitalistic system as central economies do. Consequently the history of central capitalism is, at the same time, the history of peripheral capitalism. But specific links between dependent and central economies could have been different in each of the above periods. The same can be argued *vis-à-vis* analyses about competitive or monopolistic trends in the development of capitalism and its effects on peripheral economies.

During these different phases of the capitalist process, the Latin American countries depended on various countries that acted as centres and whose economic structures influenced the nature of the dependence. For example, Great Britain's economic expansion required some measure of development in the peripheral economies, since it relied on them to supply raw materials. Furthermore, these same economies were part of the market for its manufactured products. It was therefore necessary for Latin American production to achieve a certain degree of growth and modernization. The United States economy, on the other hand, had its own natural resources as well as a domestic market that permitted it a more autonomous development in respect of the peripheral economies; in some cases it even competed with the countries producing raw materials. The relation of dependence thus came to denote control of the development of other economies both in the production of raw materials and in the possible formation of other economic centres. The vitalizing role of the United States in the Latin American economies was therefore less important (prior to the formation of the present multinationals) than the role performed by English capitalism.

The developing countries are by no means repeating the history of the developed countries. Historical conditions are different. When the world market was created along with development, it was thanks to the action of the 'bourgeoisie conquérante'. Now, development is undertaken when capitalist market relations already exist between both groups of countries and when the world market is divided between the capitalist and socialist worlds. What at first glance may appear to be deviant forms of the classic development pattern are simply not. When we recognize this, the present socio-economic system in dependent countries may become understandable. (. . .)

The more developed countries of Latin America are attempting to define foreign policy objectives that take advantage of contradictions in the international order and allow these countries some independent policy-making. But these countries remain dependent and assure an internal social order favourable to capitalist interests and consequently fail to challenge one of the basic objectives of American foreign policy. Multinational enterprises continue to receive support from the foreign policies of their countries of origin, as well as from local states.

How can these contradictory forces act together? It is through contradictions that the historical process unfolds. Dependent development occurs through frictions, accords, and alliances between the state and business enterprises. But this type of development also occurs because both the state and business enterprises pursue policies that form markets based on the concentration of incomes and on the social exclusion of majorities. These processes demand a basic unity between these two historical actors as they confront popular opposition, which may be activated when nationalist or socialist movements question the existing social order. So, the conflicts between the state and Big Business are not as antagonistic as the contradictions between dominant classes and people.

Within the last ten years, the strengthening of the state and the penetration of multinational corporations occurred within the context of a new set of class relations. On one hand, attempts are made to break (sometimes radically) with the global situation of dependency, with the aim of transforming society in the direction of socialism. On the other hand, dominant classes were re-ordered, with emphasis placed on the repressive role of the state and on the simultaneous transformation of the state into a tool for the fortification of the capitalist economic order.

The exhaustion of the prior populism and the aggravation of class tensions gave rise to various political attempts to break with the prevailing style of development. In one form or another, during the past decade, the politics of Latin American popular forces were profoundly marked by the presence of the Cuban revolution. The shadow of Guevara's deeds and the quasi-substitution of the process of mass politics by the military actions of guerrilla groups (though this was not implicit in their theory) considerably polarized Latin-American revolutionary movements. These attempts failed nearly everywhere, the only exception of consequence being the case of Argentina, where the two principal guerrilla currents were not completely dissociated from the remaining socio-political movements. Though not constituting a real political power alternative, the guerrillas of Argentina exert a certain veto capacity, conditioning other political movements and attempts at reformulating class alliances.

Attempts at radical rupture with the capitalist-developmentalist path were not limited to the politics of the guerrilla. The Chilean popular unity of the Allende period, as one case, and the Peruvian military reformism, as another, were reactions based on broader popular forces to development that is tied to international capitalist-oligopolistic expansion. In both cases the state was viewed not as a 'bourgeois institution' to be destroyed, but as the lever for a possible total transformation of society, on condition that its control remain in the hands of popular forces.

Both the battle between classes and the basic dependency relationship find in the state a natural crossroads. The contradiction of a state that constitutes a nation without being sovereign is the nucleus of the subject matter of dependency. Our rereading the history has proceeded throughout the book toward specifying the fundamental historical actors: classes and groups defined within specific forms of production. Now, after ten years of reasonable rates of economic growth, the expansion of global commerce, the industrialization of important segments of the periphery of the capitalist world, and the strengthening of the state productive sector, the problem unfolds in a more complex manner. *Strictu sensu*, the capacity for action of various Latin American states has increased. In this sense, one might consider that they are 'less dependent'. Our concern is not, however, to measure degrees of dependency in these terms—which fail to ask, 'less for whom? for which classes and groups?' Which classes have become more sovereign? Which alliances and class interests within each country and at the international level lead the historical process of economic development?

If the state has expanded and fortified itself, it has done so as the expression of a class situation which has incorporated both threats of rupture with the predominant pattern of capitalist development, as we have said, and policies of the dominant classes favourable to the rapid growth of the corporate system, to alliances between the state and business enterprises, and to the establishment of interconnections, at the level of the state productive system, between 'public' and multinational enterprises. To accomplish this, the state has assumed an increasingly repressive character, and dominant classes in a majority of countries have proposed policies increasingly removed from popular interest. They have rendered viable a 'peripheral' capitalist development, adopting a growth model based on replication—almost in caricature of the consumption styles and industrialization patterns of the central capitalist countries. The tendencies indicated in chapter six developed with increasing velocity, achieving successes for that style of development (the 'Brazilian miracle' and the type of growth that occurred in Mexico until 1970, are notable examples of the trend). Given conditions in Latin America, this process, while producing economic growth, urbanization, and wealth, has redefined without eliminating, or else in certain cases has aggravated the existential, social, and economic problems of a majority of the population. This majority has come to be looked upon as a resource for the accumulation of capital more than as the effective potential for the creation of a society modelled on its own interests.

Under these conditions, the state and the nation have become separated: all that is authentically popular, even if lacking the character of specific class demands, has come under suspicion, is considered subversive, and encounters a repressive response. In this vein, even problems which Western capitalist democracies confront and absorb, like the discussion of income distribution, minority movements (blacks, Indians, migrants, etc.), feminist or youth demands (not to mention the freedom of syndical and political organization), appear threatening to the existing order. From the perspective of the dominant classes, the nation has become increasingly confused with the state, and the latter in turn has identified its interests with theirs, resulting in the confusion of the public interest with the defence of the business enterprise system.

Local dominant groups in Latin America responded to the external influences on economic growth and to the need to guard against attempts to transform the prevailing order, with an amalgam between a repressive state (often under corporate military control) and an entrepreneurial state. What lends dynamism to this form of state, and what characterizes its movement, is *not* the bureaucratic aspect it may have assumed in some countries (Peru, Mexico, Brazil, Chile, among the most characteristic cases), but rather its *entrepreneurial* aspect, which leads it to ally itself, in production, with the multinational corporation. Somehow, the state has become a strategic element, functioning as a hinge that permits the opening of the portals through which capitalism passes into industrializing peripheral economies.

A state which expanded the public sector *at the same time* that it intensified relations between the latter and the multinational corporations began to develop with the accords on the 'Chileanization' of copper proposed by the government of Frei. The proposal was uncommon in the statist tradition of Latin America: the connection with foreign enterprises would be made through their association, not with the local bourgeoisie, but with public enterprises created by the state, which come to function as *corporations*.

The generalization of this model, in Brazil, in Mexico, in Peru, in Venezuela, for example, transferred the conflicts *among associates* to a more directly political sphere. In addition, it married foreign interests with the local bourgeoisie, and in certain countries, with the interests of local states in so far as they were direct agents of production, as occurred in Brazil, in Mexico, and to a lesser extent in Venezuela. The consequences of this process are enormous and are far from having been exhausted by historical practice or by analysis. The character of this state-as-entrepreneur and of the state associated economically with imperialist forces without being a politically associated state has lent to the contemporary form of the state a significance different from that which it had until mid-1950.

What is novel is the expansion of the state's direct productive investment in capitalistically profitable sectors. While state investments in these sectors originally came about with resources obtained through taxes and duties, they subsequently reproduced and expanded through the *profits* generated by the state enterprises (petrochemicals, mining, direct consumer goods, etc.). In countries like Brazil, Chile, Colombia, Peru, Mexico, and Venezuela, the public sector contributes more than 50 per cent to the annual formation of capital, with the remainder contributed by private national and foreign enterprises. Of this total, in a majority of these countries, the *state enterprises* (as an individual portion of public expenditure) constitute more than half of the investment of the public sector. In Brazil, in 1975, this figure exceeded 30 per cent of the total investment (public and private). Also in Brazil, the only two local enterprises which, by the scope of their action, could hope to qualify as multinationals (aside from the Itaipu hydroelectric corporation) are state enterprises: the Vale do Rio Doce and Petrobras. Counted among the largest enterprises operating in Brazil, in terms of assets and the value of production or trade (and leaving foreign enterprises aside), are not the enterprises controlled by local private capital, but rather those of the state. In 1975 fifty-six of the one hundred largest Brazilian enterprises were state owned.[2]

The role of bureaucracies and of technocrats is considerable in practically all of the industrialized countries of Latin America. In a penetrating essay on this subject,[3] Guillermo O'Donnell attempts to show the nature of this form of regime and the

---

[2] It should be made clear that despite the importance of the role of the state productive sector in the Brazilian economy, foreign enterprises control between 40 and 50 per cent of the large groups, according to measures of fixed assets, liquid assets, employment, and invoicing.

[3] O'Donnell, Guillermo, 'Reflexiones sobre las tendencias generaies de cambio en el Estado burocratico autoritario' (Buenos Aires, CEDES, 1975).

conditions under which it emerges. He points out that regimes of this type established themselves in the region as the response of local dominant classes to the challenge presented by the mobilization and popular pressure generated by the collapse of previous political orders (either populist or traditionally authoritarian). He adduces further that this collapse occurred when economic difficulties that followed the import-substitution stage of industrialization created an inflationary situation and led the economy into an impasse. Its solution required, aside from stability to ensure economic predictability, additional capital flows and greater entrepreneurial centralization in order to proceed along an oligopolistic route toward the continuation of the process of accumulation and toward the development of productive forces. O'Donnell concludes that, for all of these reasons, there exists a relation of 'mutual indispensibility' between bureaucratic-authoritarian states and international capital (which needs to penetrate local economies and which possesses the technological and financial requisites to undertake the 'deepening of development').

The lack of local private investment potential, the political need to prevent multinational corporations from singlehandedly appropriating the most strategic sectors of the economy and their most dynamic branches, and even, at times, the non-existence of international capital flows to attend to the investment needs of peripheral countries during any given period (since multinationals act on a global scale, aiming at maximizing results and not toward the continuity of local development), has led local states, despite the capitalist ideology they defend, to expand their functions and thereby to create a national basis from which to bargain with the multinationals. In this process, neither the decisions of the state nor the pressure from multinationals excludes local enterprises from the game. But in practice these local enterprises continue to lag behind the principal agents of transformation: the multinationals and the state. By the very force of expansion, new investment prospects do at times open up for segments of the local bourgeois sectors. Some of these return to the political-economic offensive, often allying themselves with the multinational enterprises in the 'anti-statist' struggle.

FURTHER READING

*

Cohen, G. A. *Karl Marx's Theory of History: A Defence* (1978)
Luxemburg, Rosa *The Russian Revolution* (1961)
Moore, Barrington *Social Origins of Dictatorship and Democracy* (1966)
Shaw, William H. *Marx's Theory of History* (1978)
Sweezy, Paul M. *et al. The Transition from Feudalism to Capitalism* (1976)

PART VII
*
Modern Capitalism and Imperialism

PRESENTATION

*

Marx's analysis of capitalism, which remained incomplete at his death, has been extended and revised by Marxist thinkers to take account of the main tendencies in the subsequent development of the capitalist mode of production. Five broad approaches can be distinguished in this later work, all of which, however, recognize as major characteristics of twentieth-century capitalism: first, the economic dominance of large corporations and the associated decline of competitive capitalism, and second, the growth of state intervention to regulate the economy as a whole.

The first of these reconstructions of Marxist theory is to be found in Hilferding's conception of 'organized capitalism', which he adumbrated in *Finance Capital* (1910. English trans. 1981) and elaborated in several articles published in the 1920s (see the text below). Hilferding emphasized very strongly the two features just mentioned—large corporations and state intervention—and argued that 'organized capitalism', by its extensive socialization of the economy, created the economic basis for a direct transition to socialism, although this could only be finally accomplished by means of a political struggle.

Hilferding's theory profoundly influenced Bukharin, who nevertheless criticized it for exaggerating the degree of 'stabilization' of capitalism, and for replacing the idea of economic breakdown and a revolutionary transformation of society by the more 'reformist' conception of a gradual transition to socialism on the basis of an economy already largely socialized by capitalism. In opposition to this view Bukharin (1926), and also Varga (1935), characterized the new stage of capitalist development in the 1920s and 1930s as 'state capitalism', thus emphasizing particularly the regulating activity of the state, but at the same time insisting that this had created only a 'relative stabilization' and could not eliminate serious economic crises or in the extreme case a breakdown.

A third, more recent, conception of modern capitalism is that propounded by Baran and Sweezy (1966), who direct attention primarily to the changes brought about by monopoly. This view is succinctly formulated by Sweezy in an essay (1972) where he argues that '. . . competition inevitably gives way to monopoly via the concentration and centralization of capital, and monopoly retards the accumulation process, giving rise to ever more powerful tendencies to stagnation.' From this standpoint the main determinant of a transition to socialism is the need to overcome stagnation by a rational and planned use of the growing surplus.

During the past two decades another theoretical approach has been developed, notably by Marxist thinkers in the German Democratic Republic, which characterizes the present stage as that of 'state monopoly capitalism'.[1] This emphasizes once again the important economic role of the state, which is broadly interpreted as weakening the tendency to economic collapse and as creating the material basis for socialism by the increasing socialization of production.[2]

Finally, there are diverse theoretical conceptions of 'late' or 'advanced' capitalism which likewise concentrate to a large extent upon the role of the state in regulating and planning the economy, but differ in their analyses of the relation between the state and the multinational corporations (see the text by Mandel, below), or in their assessment of the major elements of crisis and the principal contending forces in modern capitalist society (see the text by Habermas below).

The theoretical views we have sketched, notwithstanding important differences among them, do seem to converge, in some degree, upon the idea of 'organized capitalism'[3] as a new stage of development which is characterized by more extensive planning and greater stability of the economy. What is now at issue is the nature of a possible transition to socialism in these new conditions; and this is linked with two major problems, concerning

[1] This conception has been relatively little discussed in English, but there is a comprehensive account in Paul Wenlock, 'The Theory of State Monopoly Capitalism' (Unpublished Ph.D. thesis, University of Leeds 1981), and a brief discussion in Hardach and Karras (1974), ch. 4, where some other theories of modern capitalism are also examined.

[2] There are obvious affinities with Hilferding's theory of 'organized capitalism', but the political implications of such a conception are not elaborated and the theory of 'state monopoly capitalism' remains wedded, however incongruously, to a more or less orthodox Leninist view of the transition from capitalism to socialism.

[3] This has become a salient feature of much recent Marxist analysis; see, for example, the later work of Lucien Goldmann, and especially his 'Reflections on history and class consciousness', in Istvan Meszaros (1971), pp. 65–84.

on one side the political role of classes and other social groups in present day capitalist societies (see Part III above), and on the other side the difficulties encountered in the effective management of socialist economies. There is a further important question, posed most clearly by Hilferding in his last writings,[4] about the consequences of the growth in the independent power of the state, which carries within itself the germs of a totalitarian regime (see Part IV above).

The debate about the nature of capitalism in its most recent stage of development has also engendered some fundamental methodological disagreements. Thus Mandel (1975) argues that the 'abstract' laws of capitalism discovered by Marx are still operative in this new stage, and that it is possible to derive its characteristics directly from these laws. On the other hand, Uno (see the text below) makes a sharp distinction between a 'pure' theory and a 'stages' theory of capitalism, and argues that the latter, which has to take account of a variety of historical influences, leads on to an empirical analysis of capitalism, either on a world scale or in a particular country. This approach is quite close to that which has been taken by many recent Marxist thinkers in reconsidering the general relation between mode of production and social formation (see Part II above), and it opens up the prospect of more thorough and illuminating studies of particular capitalist nations, which would contribute significantly to a more adequate general analysis of present day capitalism.

[4] See especially, 'State Capitalism or Totalitarian State Economy' (1940). Reprinted in *Modern Review* (New York), I, 1947; and *Das historische Problem* (1941). English trans. of part in Tom Bottomore (1981), pp. 125–37.

KARL MARX

*

*The Capitalist Process of Production**

We have seen that the capitalist process of production is a historically determined form of the social process of production in general. This process is, on the one hand, a process by which the material requirements of human life are produced, and on the other hand, a process which takes place under specific historical and economic conditions of production and which produces and reproduces these conditions of production themselves, and with them the human agents of this process, their material conditions of existence and their mutual relations, that is, their particular economic form of society. For the aggregate of the relations in which the agents of this production stand to Nature and to each other, and within which they produce, is precisely society, considered from the point of view of its economic structure. Like all its predecessors, the capitalist process of production develops under definite material conditions which are at the same time the bearers of definite social relations into which individuals enter in the process of producing their life's requirements. These conditions and relations are, on the one hand, prerequisites, on the other hand, results and creations, of the capitalist process of production. They are produced and reproduced by it. We have also seen that capital (the capitalist is merely capital personified and functions in the process of production as the agent of capital), in the social process of production corresponding to it, pumps a certain quantity of surplus labour out of the direct producer, the worker, surplus labour for which no equivalent is returned and which always remains essentially forced labour, no matter how much it may seem to be the result of a freely concluded contract. This surplus labour is represented by a surplus value, and this surplus value is embodied in a surplus product. Surplus labour generally, in the sense of a quantity of labour beyond that required to satisfy existing needs, there must always be. But in the

* From Karl Marx, *Capital*, vol. iii, Chapter 48. Translated by Tom Bottomore.

capitalist system as in the slave system, etc., it has an antagonistic form and is complemented by the complete idleness of a section of society. A certain quantity of surplus labour is required in order to meet various contingencies, as well as for the necessary, progressive expansion of the process of reproduction (called accumulation from the point of view of the capitalist) in accordance with the development of needs and the increase of population. It is one of the civilizing aspects of capital that it imposes this surplus labour in a manner and under conditions which are more favourable to the development of the productive forces, and of social relations, and to the creation of the elements for a new and higher social structure than was the case in the preceding forms of slavery, serfdom, etc. Thus it leads on the one hand to a stage in which coercion and the monopolization of social development (including its material and intellectual advantages) by one section of society at the expense of the other section are eliminated; on the other hand it creates the material requirements and the germ of conditions which, in a higher form of society, make it possible to combine this surplus labour with a greater reduction of the time devoted to material labour. For, according to the development of the productive power of labour, the amount of surplus labour may be large in a short working day and relatively small in a long working day.

KARL MARX

*

*Knowledge and the Production Process**

The exchange of living labour for objectified labour, i.e. the establishment of social labour in the form of the antithesis between wage labour and capital, is the ultimate development of the *value relation* and of production based upon value. Its presupposition is, and remains, the mass of direct labour time, the quantity of labour utilized, as the decisive factor in the production of wealth. But to the extent that large scale industry develops, the creation of real wealth becomes less dependent on

* From Karl Marx, *Grundrisse der Kritik der politischen Ökonomie* (1953 edn.), pp. 592–4. Translated by Tom Bottomore.

labour time and on the amount of labour used than on the power of the agents which are set in motion during labour time. Their 'powerful effectiveness', in turn, is itself quite unrelated to the direct labour time required for their production, but depends far more upon the general state of science and the progress of technology, or the application of this science to production. (The development of this science, particularly natural science, and other sciences along with it, is itself related to the development of material production.) Agriculture, for example, becomes simply the application of the science of material metabolism, and the most advantageous way of using it to the greatest benefit of the whole social body. Real wealth manifests itself rather—as large scale industry reveals—in the enormous disproportion between the labour time utilized and its product, as well as in the qualitative disproportion between labour, reduced to a pure abstraction, and the power of the production which it supervises. Labour no longer seems to be incorporated to the same extent in the production process; rather, the human being acts as the watchman and regulator of the production process itself. (This applies not only to machinery, but to the co-ordination of human activities and the development of human intercourse.) The worker no longer inserts a modified natural object between the material and himself, instead, he inserts the natural process, which he transforms into an industrial one, as a means between himself and inorganic nature which he thereby masters. He takes his place alongside the production process, instead of being its principal agent. In this transformation, what appears as the great mainstay of production and wealth is neither the direct labour which the human being himself performs, nor the time during which he works, but the appropriation of his own general productive power, his understanding and mastery of nature by virtue of his existence as a social body—in short, the development of the social individual. The *theft of alien labour time, on which present day wealth is based*, appears a miserable foundation compared with this newly-developed one, created by large-scale industry itself. As soon as labour in its direct form has ceased to be the great source of wealth, so labour time ceases and must cease to be the measure of wealth, and hence exchange value [must cease to be the measure] of use value. The *surplus labour of the masses* has ceased to be the condition for the development of general wealth, just as the *non-labour of the few* has ceased to be the condition for the development of the general powers of the human head. Production based upon exchange value then breaks down, and the direct, material production process is stripped of its impoverished, antithetical form. The free development of individuals, and therefore not the reduction of necessary labour time in order to breed surplus labour, but a general reduction of the necessary labour of society to a minimum, which has as its counterpart the artistic, scientific, etc. cultivation of individuals in the free time, and with the means created, which are available to all. Capital itself is the contradiction in action [in] that it strives to reduce labour time to a minimum, while on the other hand it establishes labour time as the sole measure and source of wealth. Thus it reduces labour time in its necessary form, in order to increase it in its superfluous form, and hence establishes the superfluous, to an increasing extent, as a condition—a matter of life and death—for the necessary. On one side, then, it brings to life all the powers of science and nature, as well as those of social co-ordination and social intercourse, in order to make the creation of wealth (relatively) independent of the labour time devoted to it. On the other side, it wants to use labour time as a measure for the giant social forces thus created, and to confine them within the limits which are required in order to maintain the value already created as value. Productive forces and social relations—two different sides of the development of the social individual—appear to capital only as a means, and are merely a means, to produce from its own restricted base. In fact, however, they are the material conditions for blowing up this base. 'A nation is truly rich when, instead of working twelve hours, it works only six. Wealth is not command over surplus labour time [real wealth] but disposable time, beyond that used in immediate production, for each individual and for the whole of society.'[1]

Nature constructs no machines, no locomotives, railways, electric telegraphs, self-acting mules, etc. They are products of human industry, natural materials transformed into instruments of the human domination of nature, or its manifestation in nature. They are *instruments of the human brain, created by the human hand*; the objectified power of knowledge. The development of fixed capital indicates the extent to which general social

---

[1] Cited by Marx from *The Source and the Remedy of the National Difficulties* (1821) (anon.), p. 6 [Eds.].

knowledge has become a *direct force of production*, and thus the extent to which the conditions of the social life process have been brought under the control of the general intellect and reconstructed in accordance with it. It shows to what degree the social forces of production are produced, not only in the form of knowledge, but also as direct instruments of social practice, of the real life process.

FRIEDRICH ENGELS
*
Crises, Joint-Stock Companies, and State Intervention*

We have seen that the ever-increasing perfectibility of modern machinery is, by the anarchy of social production, turned into a compulsory law that forces the individual industrial capitalist always to improve his machinery, always to increase its productive power. The bare possibility of extending the field of production is transformed for him into a similar compulsory law. The enormous expansive force of modern industry, compared with which that of gases is mere child's play, appears to us now as a *necessity* for expansion, both qualitative and quantitative, that laughs at all resistance. Such resistance is offered by consumption, by sales, by the markets for the products of modern industry. But the capacity for extension, extensive and intensive, of the markets is primarily governed by quite different laws that work much less energetically. The extension of the markets cannot keep pace with the extension of production. The collision becomes inevitable, and as this cannot produce any real solution so long as it does not break in pieces the capitalist mode of production, the collisions become periodic. Capitalist production has begotten another 'vicious circle'.

As a matter of fact, since 1825, when the first general crisis broke out, the whole industrial and commercial world, production and exchange among all civilized peoples and their more or

---

* From Friedrich Engels, *Socialism: Utopian and Scientific* (1880). Text of the authorized English edition (1892) with minor revisions.

less barbaric hangers-on, are thrown out of joint about once every ten years. Commerce is at a standstill, the markets are glutted, products accumulate, as multitudinous as they are unsaleable, hard cash disappears, credit vanishes, factories are closed, the mass of the workers are in want of the means of subsistence, because they have produced too much of the means of subsistence; bankruptcy follows upon bankruptcy, forced sale upon forced sale. The stagnation lasts for years; productive forces and products are wasted and destroyed wholesale, until the accumulated mass of commodities finally filters off, more or less depreciated in value, until production and exchange gradually begin to move again. Little by little the pace quickens. It becomes a trot. The industrial trot breaks into a canter, the canter in turn grows into the headlong gallop of a perfect steeplechase of industry, commercial credit, and speculation which finally, after breakneck leaps, ends where it began—in the ditch of a crisis. And so on, over and over again. We have now, since the year 1825, gone through this five times, and at the present moment (1877) we are going through it for the sixth time. And the character of these crises is so clearly defined that Fourier hit all of them off when he described the first as *crise pléthorique*, a crisis from plethora.

In these crises, the contradiction between socialized production and capitalist appropriation ends in a violent explosion. The circulation of commodities is, for the time being, stopped. Money, the means of circulation, becomes a hindrance to circulation. All the laws of production and circulation of commodities are turned upside down. The economic collision has reached its apogee. *The mode of production is in rebellion against the mode of exchange.*

The fact that the socialized organization of production within the factory has developed so far that it has become incompatible with the anarchy of production in society, which exists side by side with and dominates it, is brought home to the capitalists themselves by the violent concentration of capital that occurs during crises, through the ruin of many large, and a still greater number of small, capitalists. The whole mechanism of the capitalist mode of production breaks down under the pressure of the productive forces, its own creations. It is no longer able to turn all this mass of means of production into capital. They lie fallow, and for that very reason the industrial reserve army must also lie fallow. Means of production, means of subsistence, available labourers, all the elements of production and of general wealth, are present in abundance. But 'abundance becomes the source of distress and want' (Fourier), because it is the very thing that prevents the transformation of the means of production and subsistence into capital. For in capitalistic society the means of production can only function when they have undergone a preliminary transformation into capital, into the means of exploiting human labour power. The necessity of this transformation into capital of the means of production and subsistence stands like a ghost between these and the workers. It alone prevents the coming together of the material and personal levers of production; it alone forbids the means of production to function, the workers to work and live. On the one hand, therefore, the capitalistic mode of production stands convicted of its own incapacity to further direct these productive forces. On the other, these productive forces themselves, with increasing energy, press forward to the removal of the existing contradiction, to the abolition of their quality as capital, to the *practical recognition of their character as social productive forces*.

This rebellion of the productive forces, as they grow more and more powerful, against their quality as capital, this stronger and stronger command that their social character shall be recognized, forces the capitalist class itself to treat them more and more as social productive forces, so far as this is possible under capitalist conditions. The period of industrial high pressure, with its unbounded inflation of credit, no less than the crash itself, by the collapse of great capitalist establishments, tends to bring about that form of the socialization of great masses of means of production which we meet with in the different kinds of jointstock companies. Many of these means of production and of distribution are, from the outset, so colossal that, like the railways, they exclude all other forms of capitalistic exploitation. At a further stage of evolution this form also becomes insufficient. The producers on a large scale in a particular branch of industry in a particular country unite in a trust, a union for the purpose of regulating production. They determine the total amount to be produced, parcel it out among themselves, and thus enforce the selling price fixed beforehand. But trusts of this kind, as soon as business becomes bad, are generally liable to break up, and on this very account compel a yet greater concentration of association. The whole of the particular industry is turned into one gigantic joint-stock company; internal competition gives place to the internal monopoly of this one company. This has happened in 1890 with the English alkali production, which is now, after the fusion of 48 large works, in the hands of one company, conducted upon a single plan, and with a capital of £6,000,000.

In the trusts, freedom of competition changes into its very opposite—into monopoly; and the production without any definite plan of capitalistic society capitulates to the production upon a definite plan of the invading socialistic society. Certainly this is so far still to the benefit and advantage of the capitalists. But in this case the exploitation is so palpable that it must break down. No nation will put up with production conducted by trusts, with so barefaced an exploitation of the community by a small band of coupon clippers.

In any case, with trusts or without, the official representative of capitalist society—the state—will ultimately have to undertake the direction of production.[1] This necessity for conversion into state property is felt first in the great institutions for intercourse and communication—the post office, the telegraphs, the railways.

If the crises demonstrate the incapacity of the bourgeoisie for

---

[1] I say '*have to*'. For only when the means of production and distribution have *actually* outgrown the form of management by joint-stock companies, and when, therefore, the taking them over by the state has become *economically* inevitable, only then—even if it is the state of today that effects this—is there an economic advance, the attainment of another step preliminary to the taking over of all productive forces by society itself. But of late, since Bismarck went in for state ownership of industrial establishments, a kind of spurious socialism has arisen, degenerating, now and again, into something of flunkeyism, that without more ado declares *all* state ownership, even of the Bismarckian sort, to be socialistic. Certainly, if the taking over by the state of the tobacco industry is socialistic, then Napoleon and Metternich must be numbered among the founders of socialism. If the Belgian state, for quite ordinary political and financial reasons, itself constructed its chief railway lines; if Bismarck, not under any economic compulsion, took over for the state the chief Prussian lines, simply to be the better able to have them in hand in case of war, to bring up the railway employees as voting cattle for the government, and especially to create for himself a new source of income independent of parliamentary votes—this was, in no sense, a socialistic measure, directly or indirectly, consciously or unconsciously. Otherwise, the Royal Maritime Company, the Royal porcelain manufacture, and even the regimental tailor shops of the Army would also be socialistic institutions, or even, as was seriously proposed by a sly dog in Frederick William III's reign, the taking over by the state of the brothels.

managing any longer modern productive forces, the transformation of the great establishments for production and distribution into joint-stock companies, trusts, and state property shows how unnecessary the bourgeoisie are for that purpose. All the social functions of the capitalist are now performed by salaried employees. The capitalist has no further social function than that of pocketing dividends, tearing off coupons, and gambling on the Stock Exchange, where the different capitalists despoil one another of their capital. At first the capitalistic mode of production forces out the workers. Now it forces out the capitalists, and reduces them, just as it reduced the workers, to the ranks of the surplus population, although not immediately into those of the industrial reserve army.

But the transformation, either into joint-stock companies and trusts, or into state ownership, does not do away with the capitalistic nature of the productive forces. In the joint-stock companies and trusts this is obvious. And the modern state, again, is only the organization that bourgeois society takes on in order to support the external conditions of the capitalist mode of production against the encroachments as well of the workers as of individual capitalists. The modern state, no matter what its form, is essentially a capitalist machine, the state of the capitalists, the ideal personification of the total national capital. The more it proceeds to the taking over of productive forces, the more does it actually become the national capitalist, the more citizens does it exploit. The workers remain wage-workers—proletarians. The capitalist relation is not done away with. It is rather brought to a head. But, brought to a head, it topples over. State ownership of the productive forces is not the solution of the conflict, but concealed within it are the technical conditions that form the elements of that solution.

This solution can only consist in the practical recognition of the social nature of the modern forces of production, and therefore in the harmonizing of the modes of production, appropriation, and exchange with the socialized character of the means of production. And this can only come about by society openly and directly taking possession of the productive forces which have outgrown all control except that of society as a whole. The social character of the means of production and of the products today reacts against the producers, periodically disrupts all production and exchange, acts only like a law of Nature working blindly, forcibly, destructively. But with the taking over by society of the productive forces, the social character of the means of production and of the products will be utilized by the producers with a perfect understanding of its nature, and instead of being a source of disturbance and periodical collapse, will become the most powerful lever of production itself.

RUDOLF HILFERDING

*

*The Organized Economy**

If we want to know what the present situation really is we must examine it much more closely and characterize it more precisely than is done by the phrase 'late capitalism'. The crucial point is that we find ourselves at present in the period of capitalism in which the era of free competition, during which capitalism was wholly under the sway of blind market laws, has been essentially superseded, and we are moving towards a capitalist organization of the economy; in short, from an *economy regulated by the free play of forces to an organized economy.*

The technological characteristics of the organized economy—briefly summarized—are that alongside steam and electrical power, synthetic chemicals now play an increasingly prominent role, after something like half a century of scientific development during which they were becoming ripe for technical application in manufacturing. This application of chemistry signifies something quite new in principle. It makes the capitalist economy independent of the supply of individual raw materials, since the aim is, in principle, to produce important raw materials artificially from inorganic substances which are everywhere readily available in large quantities. (. . .) A second aim of synthetic chemistry is to produce raw materials in such a form that they are far more suitable than natural materials for industrial use, or have quite new qualities. A third aim of this development is to produce costly organic materials out of cheap

* From Rudolf Hilferding, 'Die Aufgaben der Sozialdemokratie in der Republik' (1927). Translated by Patrick Goode. Published by permission of Dr Peter Milford.

inorganic materials. Here I need only recall the enormous progress of artificial silk production which has made such inroads into the former realm of the textile industry. (. . .)

The second characteristic of the situation is that capitalist industry, reanimated by the vigorous and unprecedented influx of scientific knowledge, is wholly committed to utilizing the new opportunities in an *organized* way. It is significant that newly-established industries are not only built upon a more complex technological base (as was already the case in the immediately preceding period) but at the same time are organizing themselves, so far as possible, on a world-wide scale. For example, the artificial silk industry is not only a monopoly industry in Germany, but fundamentally constitutes a single international capitalist concern which has close links with other trusts in Germany and England, and these in turn have connections with other concerns. Thus the development of *cartels* and *trusts* which has been successfully accomplished in industry is now the first word to be uttered by the new industries as they enter upon the world scene.

A third characteristic phenomenon is the *internationalization* of capitalist industry; the effort to unite the various national monopolies, trusts, and cartels at an international level. Anyone who comes into contact with capitalist business circles (. . .) is astonished at the eagerness with which these people, whose economic outlook before the war was confined to the national scene, now seek international connections and cultivate their foreign relations, and how vigorous this trend towards international organization has become. When the working class was first becoming organized and the trade unions formed the first organized economic element in capitalism, the entrepreneurs, helped by their greater class-consciousness and their smaller numbers, began to overtake our organizations. We must take care that this does not also happen in the international arena. (. . .)

The fourth point, which is not usually noticed and is only just beginning to be apparent is perhaps the most important of all. We all have the feeling nowadays that even those businesses carried on by a single entrepreneur have ceased to be simply a private affair of the entrepreneurs. Society has come to realize that it has an interest in raising productivity in every individual undertaking, and hence that the person responsible for economic management in each case should actually carry out his technical and organizing functions to increase productivity. I should remind you that bodies such as the 'Kuratorium für Wirtschaftlichkeit', and all those officially promoted efforts for greater rationalization which are designed to induce entrepreneurs to increase the output of their undertakings, are simply statements by society to the effect that the running of his business is no longer the private affair of the entrepreneur but a matter for society as a whole. The most important thing is this: the formation of combines, the concentration of increasing numbers of undertakings under a single head, means the elimination of competition so far as the individual business is concerned. It was the doctrine of capitalism that only the pressure of free competition can stimulate the economy and bring about the necessary technical innovations and advances. The principal argument against socialism has always been: you eliminate the private initiative of free competition and offer nothing in its place. Hence your economy will not work, because you fail to take into account the ambition and self-interest of the private owner of the means of production. It is most interesting to see how, in the development of modern business science, methods are being sought of replacing this free competition between private self-interests with scientific methods of planning. It is quite clear that the director of a combine has a great interest in being able to establish at any time whether any one company which forms part of his concern, but is not in competition with other similar undertakings in the combine, is operating at maximum efficiency. Very sophisticated methods have been developed in order to substitute for competition based on self-interest, a *scientific method of competition*. That is the very principle which we socialists would apply to our management of the economy. Capitalism itself thus abandons the principal objection which it can raise against socialism, and at the same time the last psychological objection to socialism collapses. If fact, therefore, organized capitalism means the *theoretical replacement of the capitalist principle of free competition* by the *socialist principle of planned production*. This planned and consciously directed economy supports to a much greater extent the possibility of the conscious action of society; that is to say, action by means of the only conscious organization in society, equipped with coercive power, namely the state.

If this is the case, then the capitalist organization of the economy on one side, and the state organization on the other, confront each other, and the problem is how we want to shape their interpenetration. That simply means that our generation faces the problem of transforming—with the help of the state, which consciously regulates society—an economy organized and directed by the *capitalists* into one which is directed by the *democratic state*. It follows that the problem facing our generation can be none other than that of socialism. If we, as Social Democracy, fought in an earlier period for political rights, for the initiation and extension of social policies, then as a result of economic development itself the *question of socialism* is now posed. (. . .)

I referred to the growing interpenetration of economy and state, of their reciprocal relation which becomes increasingly close as the economy becomes more organized. In this connection I draw attention to the fact that even during the period of free competition the influence of the state upon the economy already existed in certain spheres; for example, in the state control of the money market which has become very significant again in the last few days through the fact, unique in the history of stock market crises, that a capitalist government artificially produced a panic on the stock market, or in questions of fiscal policy and trade policy. Here, however, I have the impression that it is necessary to remind the mass of the people of the significance of trade policy. We have recently experienced an extraordinary rise in the price of cereals and it must be made clear to the people that the *price of bread* and the *price of meat* is not only an economic but also a *political price*, which is determined by political power relations, and that it is a matter of urgency, if the people desire an improvement in this sphere, that they take the initiative themselves in implementing and supporting a policy capable of diminishing or eliminating this political factor in the economic price.

But what is new, and even more important, is state regulation in that area which directly affects the lot of the proletariat; namely, the labour market. Thanks to the revolution we now have unemployment insurance. This means a very specific regulation of supply and demand on the labour market. As a result of the system of wage agreements and courts of arbitration there is now a *political regulation of wages* and of working hours. The personal situation of the worker is determined by the policy carried out by the state. If we have been able, with more than two million unemployed, to maintain, on the whole, the level of real wages, this defence of the wage level has been possible above all because the political influence of the working class was great enough—through unemployment insurance, arbitration, and wage agreements—to prevent at least, a reduction of wages. We must drum it into the head of every worker that *the weekly wage is a political wage*, that the size of the pay packet at the end of the week depends upon the strength of working-class representation in parliament, the strength of its organization, and the power relations in society outside parliament. And to working women in particular it must be said: When you vote, make your decision about bread and meat, and about the level of wages, at the same time. Of course, this is something new in the capitalist economy, an element of great economic, social, and political significance. Professor Cassel, that fossil from the *laisser-faire* era of capitalism, who strangely enough is able to travel around as an international expert on the subject, is quite right when he says that this contradicts the nature of capitalism—that is, capitalism as he understands it. It is indeed incompatible with the principle of free competition. It is only possible because we have an organized economy, which is subject to ever-increasing conscious organization by society and by the state.

At this point I come to our position with regard to the state. Here I should like to invoke the best of all Marxists—history—which on this occasion too is in agreement with Karl Marx. What was our attitude to the state *historically*? There is no doubt that from the very beginning the labour movement, and in particular the socialist movement, has been the bearer of the idea of state influence on the economy, in opposition to liberalism. There is no doubt that—initially in the field of social policy—we have repeatedly demanded state intervention and an increase in state power, and that we now demand its extension from the sphere of social policy to that of economic policy and the management of the economy. To regard the management of industry and the economy as social concerns, this is the very principle of socialism, and society has no other organ through which it can consciously act than the *state*. So for the present no doubts regarding our position on the state are possible. But if that is the case historically, we have always been careful to avoid lapsing into the conceptions of the bourgeois, and particularly the German philosophy of the state. Marxist method requires that in dealing with all social phenomena we should dissolve the fetishism of appearance by an analysis of reality. The German philosophy of the state has absolutized and deified the state; it has taught that the state is the realization of freedom, morality, or some other metaphysical principle. The German philosophy of the state flourished all the more profusely the less state power we actually had. Only since 1870 have we had what can be called a state, and our philosophy of the state, which dates from the 18th and the beginning of the 19th centuries, is therefore unusable in understanding it. Marx undoubtedly indicated a crucial feature of the state when he said that it is not to be regarded only as a political body, but also in terms of its social content, which consists in the fact that the ruling class maintains its domination by means of state power. But this definition of the state by Marx is on that account not a theory of the state, because it is valid for all state formations since the beginning of class society, whereas it is a matter of explaining the distinctive features in the development of the state.

The British, who have had a state for such a long time already, have never concerned themselves with mere philosophical conceptions of the state. The British literature on the subject does not deal at all with the state, but with government. For us, as socialists, it should be self-evident that an organization consists of its members, the leadership, and the apparatus of administration; this means, therefore, that the state—in political terms—is nothing but the government, the machinery of administration and the citizens who compose it. In another context this means that the essential element in every modern state is the *parties*, because the individual can only make his will effective through the medium of his party. Consequently, all parties are necessary components of the state, just like the government and the administration. This also involves acknowledgement of the basis of the Marxist definition, because the party struggle is no more than a reflection of the struggle between classes, and hence the expression of class conflict.

If, therefore, the content of the struggle over the state is to gain influence over the management of the economy, only then does the full originality of an observation by Marx become clear; an observation which he considered so important that he included it not only in *Capital*, but also in his *Inaugural Address*.[1] There he speaks of the ten-hour working day and concludes: 'Hence the Ten Hours Bill was not only a great practical success; it was the victory of a principle; it was the first time that in broad daylight the political economy of the middle class succumbed to the political economy of the working class.' What this means is that the more capitalist society succumbs to the increasing influence of the working class, the more victorious is the political principle of the working class to use the state as a means for the management and the control of the economy in the interests of all.

KOZO UNO

*

*The Pure Theory and the Stages Theory of Capitalism**

The development of capitalism in England from the seventeenth and the eighteenth century to the middle of the nineteenth century reflected, indeed, the process in which the economy of the traditional 'small producers,' largely dominated by feudal and medieval social relations, evolved into a capitalist commodity-economy, i.e. the process in which the commodity-economy increasingly realized its purer and more rational form, promising a social order of equality and freedom. Thus the classical view that capitalist society was the only ideal society was not ill-founded in the perspective of that period of capitalist evolution. But the illusion was shattered after the 1820s, when it became obvious to everyone that capitalist society was not exempt from the decennial cycle of violent eruptions in the form of industrial crises. This disillusion gave rise to socialism on the

[1] Marx's Inaugural Address to the International Working Men's Association (1864) [Eds.].

* From Kozo Uno, *Principles of Political Economy: Theory of a Purely Capitalist Society*. Translated by Thomas T. Sekine (Brighton: Harvester Press, 1980), pp. xxi–xxiii, xxvi–xxvii. Reprinted by permission of Harvester Press Ltd. and Humanities Press, Inc., New Jersey.

one hand, and the vulgarization of political economy on the other. Vulgar economics quickly relinquished the scientific investigations of Classical Political Economy and turned to the apologetics of capitalism with glorified platitudes and verbiage. The scientific tradition of political economy was, however, inherited by Karl Marx (1818–83), whose *Capital* was written with the avowed objective of securing a scientific groundwork for socialism. Marx's critical method enabled political economy for the first time to view capitalism as an historical process, in which context alone the mechanism of a commodity-economy is capable to total comprehension. Political economy was finally put on the right track in its search for the theory of pure capitalism as a self-contained scientific system.

Yet even Marx could not quite predict the profound transformation of capitalism that was to occur towards the end of the nineteenth century and thereafter. In *Capital*, therefore, he took it for granted that an increasingly pure form of capitalist society would emerge with the development of capitalism. Not only was this forecast consistent with the historical experience of capitalist development through the seventeenth and the eighteenth century to Marx's own time but a pure theory designed to explain the general nature of capitalism requires the methodological hypothesis that capitalism increasingly tends to perfection. Yet the actual history of capitalism did not materialize this hypothesis. For after the 1870s the era of finance-capital evolved, at which stage a high level of capitalist development no longer required the elimination of the traditional class of 'small producers'. Thus arose a stage of capitalist development at which the actual course of history diverged from the methodological presupposition of an increasingly purer capitalist society. Faced with this reality, political economy now demands, in addition to the abstract theory of pure capitalism, a new branch of investigation which inquires into the more concrete characteristics of the stages of capitalist development. The so-called doctrine of imperialism in Marxist political economy can be reformulated as an example of the stages-theory, although the relation between the stage-theory of imperialism and the pure theory of capitalism was never clearly understood by the original proponents of the doctrine of imperialism.[1]

[1] The so-called Revisionism inspired by Eduard Bernstein (1850–1932) claimed towards the end of the nineteenth century that Marx's economic

At this point it becomes clear how the study of political economy ought to be structured. *First*, the pure theory of capitalism must pre-suppose the abstract context of a purely capitalist society made up of the three major classes of capitalists, workers, and landowners in order to account for the laws peculiar to capitalism and the dynamics of their operation. The pure theory, which corresponds to what has traditionally been called the 'principles' of political economy, in effect reconstructs a system of all the basic categories generally associated with any capitalist commodity-economy in order to show their organic interrelations. The pure theory, in other words, reproduces a theoretical capitalist society, the self-containedness of which conclusively demonstrates the ability of capitalism to form an

doctrines had already become inapplicable to the analysis of the phenomena characteristic of the contemporary stage of capitalist development. This view was founded on a shallow understanding of Marx's theories and could be easily refuted by Karl Kautsky (1854–1938) and others who represented Marxist orthodoxy at that time. But both the assailants and the defenders of Marx's economic doctrines were incorrect in assuming that economic theory propounded in *Capital* ought to be, either immediately or with minor modifications, applied to the clarification of the new empirical phenomena. Later, however, two important books, *Das Finanzkapital* (1910) by Rudolf Hilferding (1877–1941) and *Die Akkumulation des Kapitals* (1913) by Rosa Luxemburg (1870–1919), tried to work these empirical phenomena into a synthetic concept of imperialism. Of these two authors, Luxemburg left little impact on the later development of political economy because she made a serious misinterpretation of the theories of *Capital* in her book. Hilferding's contribution, on the other hand, was much praised by Lenin, whose *Imperialism* (1917) was largely based on Hilferding's notion of finance-capital. Thus Marxists in the end adopted the Hilferding-Lenin version of the doctrine of imperialism.

The formulation of this doctrine, however, poses a new problem concerning its relation to the fundamental economic theory of *Capital*, a problem which has not been solved by Marxists. The solution involves an operation no less than the overhauling and complete restatement of *Capital* as a theory of pure capitalism. If *Capital* is left as it stands, and if the doctrine of imperialism is considered merely as an appendage, the relation between the pure theory of capitalism and the stages-theory of capitalist development cannot be fully understood. If, on the other hand, the relation of the doctrine of imperialism to *Capital* is correctly interpreted, then the doctrine cannot remain as a mere economic history of capitalism in the Imperialist Age. It must define the stage of imperialism characterized by the chrematistics of finance-capital in contrast to the earlier stages of mercantilism and liberalism shaped respectively by the activities of merchant and industrial capital. This means that the stages-theory of capitalist development must characterize the three stages of mercantilism, liberalism, and imperialism as forming the 'types' of capitalist development. The method of the stages-theory is therefore clearly different from that of the pure theory.

historical society. *Second*, political economy must direct its attention to the process of capitalist development in which all economic phenomena, representing the abstract principles of capital, always appear under more concrete and stage-characteristic forms. In the actual process of capitalist development, as already mentioned, the tendency for capitalism to approximate its purest image prevailed only up to a certain point over the factors persistently resisting integration into the commodity-economy. After that point was passed, those factors gained the upper hand, frustrating the realization of pure capitalism. Actual economic phenomena always feature certain aspects that fail to be fully explained by the pure theory; those aspects became particularly important after the road to pure capitalism was blocked. Such specialized branches of political economy as commercial, industrial, agricultural, and colonial policies together with financial institutions, public finance, and *sozialpolitik* must all presuppose the stages-theory of capitalist development; for these specialized branches study the various aspects of the capitalist economy, as they appear differently at each of the three historical stages, in the light of the world-historic 'type' set by the nation leading the current development of capitalism.[2] *Finally*, with all these preparations, political

---

[2] The study of economic policies whether commercial, industrial or any other does not offer so-called policy prescriptions. For instance, the study of commercial policies really amounts to the clarification of the development of capitalism from the point of view of commerce and the appraisal of the historical need for different types of commercial policies from one stage of capitalist development to another. Thus the study of economic policies in general leads to nothing else but the stage-theoretic formulation of capitalist development, distinguishing the mercantilist, the liberal, and the imperialist policies which correspond with the typical activities of merchant, industrial, and finance capital. In the light of this historical distinction between the types of general economic policies, studies of financial institutions and public finance can pursue more detailed analysis. It must be noticed, however, that, whereas the study of financial institutions is dependent on the theory of interest in the pure theory of capitalism, the study of public finance has no comparable underpinning in the pure theory. This is because public finance belongs partly to political science which has no pure theory of its own. On the other hand, the so-called *sozialpolitik* (social welfare policies), propounded and practised in Germany towards the end of the nineteenth century, must be quite specifically related to the definite stage of capitalist development. These policies were designed to counteract the growing impact of socialism by accommodating some of the social welfare problems including unemployment within the existing framework of the

economy can apply itself to its ultimate aim, namely, the empirical analysis of the actual state of capitalism either in the world as a whole or in each different country.³

capitalist regime. They must not be confused with more traditional measures for the relief of the paupers, even though in actual practice the same policy might have been intended to kill two birds with one stone. Political economy even in this case does not design, prescribe, or advocate any policy measures; it objectively describes them in the context of capitalist development.

³ The pure theory of capitalism cannot explain in detail the whole process of an actual economy. Although there is the tendency for capitalism to become increasingly purer at a definite stage of its development (which fact alone validates the pure theory of capitalism), the actual process of capitalist development never completely eliminates the remnants of the traditional and pre-capitalist relations. The failure to eliminate them becomes in fact quite conspicuous after the approach of capitalism to its purest form is frustrated. Hence it cannot be generally concluded that the phenomena which are left theoretically unexplained are of negligible importance as if they were bound to disappear with the development of capitalism. The countries whose capitalist evolution followed that of England were in a position to import from their predecessor the results of its capitalist development, which set the example but was not necessarily duplicated elsewhere. That is the reason why empirical studies of the actual economy, whether of the world or of a nation, must always presuppose not only the laws of pure capitalism but also the stage-characteristics of capitalist development. The necessity of the stages-theory becomes obvious, when even England, which was once the leading capitalist nation, had to depart from its pursuit of pure capitalism in the face of competition with the newly emerging industrial nations such as Germany and the United States. Although political economy since the time of the Classical School had always aimed at a pure theory of capitalism, the full implication of this aim could not become apparent until, towards the end of the nineteenth century, the age of finance-capital evolved and showed in no uncertain terms the indispensability of the stages-theory. The profound transformation of capitalism made it finally clear that the pure theory of capitalism alone was not enough to fully account for the actual state of any economic process. Only with the recognition of the need for the stages-theory of capitalist development, can the exact scope and the meaning of the pure theory itself also be made unambiguous.

JÜRGEN HABERMAS

*

A Descriptive Model of Advanced Capitalism*

The expression 'organized or state-regulated capitalism' refers to two classes of phenomena, both of which can be attributed to the advanced stage of the accumulation process. It refers, on the one hand, to the process of economic concentration—the rise of national and, subsequently, of multinational corporations—and to the organization of markets for goods, capital, and labour. On the other hand, it refers to the fact that the state intervenes in the market as functional gaps develop. The spread of oligopolistic market structures certainly means the end of *competitive capitalism*. But however much companies broaden their temporal perspectives and expand control over their environments, the steering mechanism of the market remains in force as long as investment decisions are made according to criteria of company profits. Similarly, the supplementation and partial replacement of the market mechanism by state intervention marks the end of *liberal capitalism*. None the less, no matter how much the scope of the private autonomous commerce of commodity owners is administratively restricted, political planning of the allocation of scarce resources does not occur as long as the priorities of the society as a whole develop in an unplanned, nature-like manner—that is, as secondary effects of the strategies of private enterprise. In advanced capitalist societies the economic, the administrative, and the legitimation systems can be characterized, approximately and at a very general level, as follows.

The Economic System. During the sixties various authors, using the United States as an example, developed a three-sector model based on the distinction between the private and the public sectors.[1] According to the model, private production is market-oriented, one sub-sector still being regulated by competition

* From Jürgen Habermas, *Legitimation Crisis*. Translated by Thomas McCarthy (London: Heinemann, 1976), pp. 33–9. German text copyright © 1973 by Suhrkamp Verlag. Introduction and English translation copyright © 1975 by Beacon Press. Reprinted by permission of Heinemann Educational Books and Beacon Press.

[1] Reagan (1963); Shonfield (1965); Crosser (1960); Galbraith (1967); Weidenbaum (1969); and Melman (1970).

while the other is determined by the market strategies of oligopolies that tolerate a 'competitive fringe'. By contrast, in the public sector, especially in the armaments and space-travel industries, huge concerns have arisen whose investment decisions can be made almost without regard for the market. These concerns are either enterprises directly controlled by the state or private firms living on government contracts. In the monopolistic and the public sectors, capital-intensive industries predominate; in the competitive sector, labour-intensive industries predominate. In the monopolistic and public sectors, companies are faced with strong unions. In the competitive sector workers are less well organized, and wage levels are correspondingly different. In the monopolistic sector, we can observe relatively rapid advances in production. In the public sector, companies do not need to be rationalized to the same extent. In the competitive sector, they cannot be.[2]

The Administrative System. The state apparatus carries out numerous imperatives of the economic system. These can be ordered from two perspectives: by means of global planning, it regulates the economic cycle as a whole; and it creates and improves conditions for utilizing excess accumulated capital. Global planning is limited by the private autonomous disposition of the means of production (for the investment freedom of private enterprises cannot be restricted) and positively by the avoidance of instabilities. To this extent the fiscal and financial regulation of the business cycle, as well as individual measures intended to regulate investment and over-all demand—credits, price guarantees, subsidies, loans, secondary redistribution of income, government contracts guided by business-cycle policy, indirect labour-market policy, etc.—have the reactive character of avoidance strategies within the framework of a system of goals. This system is determined by a formalistically (*leerformelhaft*) demanded adjustment between competing imperatives of steady growth, stability of the currency, full employment, and balance of foreign trade.

While global planning manipulates the boundary conditions of decisions made by private enterprise in order to correct the market mechanism with respect to dysfunctional secondary effects the state actually *replaces* the market mechanism whenever

[2] O'Connor (1973).

it creates and improves conditions for the realization of capital:

— through 'strengthening the competitive capability of the nation' by organizing supranational economic blocks, securing international stratification by imperalist means, etc.;
— through unproductive government consumption (for example, armaments and space exploration);
— through guiding, in accord with structural policy, the flow of capital into sectors neglected by an autonomous market;
— through improvement of the material infrastructure (transport, education, health, recreation, urban and regional planning, housing construction, etc.);
— through improvement of the immaterial infrastructure (general promotion of science, investments in research and development, provision of patents, etc.);
— through heightening the productivity of human labor (general system of education, vocational schools, programs for training and re-education, etc.);
— through relieving the social and material costs resulting from private production (unemployment compensation, welfare, repair of ecological damage).

Improving the nation's position in the international market, government demand for unproductive commodities, and measures for guiding the flow of capital, open up or improve chances for capital investment. With all but the last of the remaining measures this is indeed a concomitant phenomenon; but the goal is to increase the productivity of labour and thereby the 'use value' of capital (through provision of collective commodities and through qualification of labour power).

The Legitimation System. With the appearance of functional weaknesses in the market and dysfunctional side effects of the steering mechanism, the basic bourgeois ideology of fair exchange collapses. Re-coupling the economic system to the political—which in a way repoliticizes the relations of production—creates an increased need for legitimation. The state apparatus no longer, as in liberal capitalism, merely secures the general conditions of production (in the sense of the prerequisites for the continued existence of the reproduction process), but is now actively engaged in it. It must, therefore—like the pre-capitalist state—be legitimated, although it can no longer rely on residues of tradition that have been undermined and worn out during the development of capitalism. Moreover, through the universalistic value-systems of bourgeois ideology, civil rights—including the right to participate in political elections—have become established; and legitimation can be dissociated from the mechanism of elections only temporarily and under extraordinary conditions. This problem is resolved through a system of formal democracy. Genuine participation of citizens in the processes of political will-formation (*politischen Willensbildungsprozessen*), that is, substantive democracy, would bring to consciousness the contradiction between administratively socialized production and the continued private appropriation and use of surplus value. In order to keep this contradiction from being thematized, then the administrative system must be sufficiently independent of legitimating will-formation.

The arrangement of formal democratic institutions and procedures permits administrative decisions to be made largely independently of specific motives of the citizens. This takes place through a legitimation process that elicits generalized motives— that is, diffuse mass loyalty—but avoids participation. This structural alteration of the bourgeois public realm (*Öffentlichkeit*) provides for application of institutions and procedures that are democratic in form, while the citizenry, in the midst of an objectively (*an sich*) political society, enjoy the status of passive citizens with only the right to withhold acclamation. Private autonomous investment decisions thus have their necessary complement in the civic privatism of the civil public.

In the structurally depoliticized public realm, the need for legitimation is reduced to two residual requirements: The first, civic privatism—that is, political abstinence combined with an orientation to career, leisure, and consumption (see Part II, Chapter 7)—promotes the expectation of suitable rewards within the system (money, leisure time, and security). This privatism is taken into account by a welfare-state substitute programme, which also incorporates elements of an achievement ideology transferred to the educational system.[3] Second, the structural depoliticization itself requires justification, which is supplied either by democratic elite theories (which go back to Schumpeter[4] and Max Weber) or by technocratic systems

---

[3] Habermas (1971), pp. 102 ff.
[4] Schumpeter (1976).

theories (which go back to the institutionalism of the twenties).[5] In the history of bourgeois social science, these theories today have a function similar to that of the classical doctrine of political economy. In earlier phases of capitalist development, the latter doctrine suggested the 'naturalness' of the capitalist economic society.

Class Structure. While the political form of the relations of production in traditional societies permitted easy identification of ruling groups, in liberal capitalism manifest domination was replaced by the politically anonymous power of civil subjects. (Of course, during economically induced social crises these anonymous powers again assumed the identifiable form of a political adversary, as can be seen in the fronts of the European labour movement.) But, while in organized capitalism the relations of production are indeed repoliticized to a certain extent, the political form of the class relationship is not thereby restored. Instead, the political anonymity of class domination is superseded by social anonymity. That is, the structures of advanced capitalism can be understood as reaction formations to endemic crisis. To ward off system crisis, advanced-capitalist societies focus all forces of social integration at the point of the structurally most probable conflict—in order all the more effectively to keep it latent.[6] At the same time, in doing so they satisfy the political demands of reformist labour parties.[7]

In this connection, the quasi-political wage structure, which depends on negotiations between companies and unions, plays a historically epoch-making role. 'Price setting' (*Machtpreisbildung*, W. Hofmann), which replaces price competition in the oligopolistic markets, has its counterpart in the labour market. Just as the great concerns quasi-administratively control price movements in their markets, so too, on the other side, they obtain quasi-political compromises with union adversaries on wage movements. In those branches of industry belonging to the monopolistic and the public sectors, which are central to economic development, the commodity called labour power receives a 'political price'. The 'wage-scale partners' [*Tarifpartner*] find a

[5] For example, Rathenau, Berle, and Means.
[6] C. Offe, 'Political Authority and Class Structure—An Analysis of Late Capitalist Societies', in *International Journal of Sociology* (Spring 1972), pp. 73–108. [See the excerpt on pp. 139–45 above. Eds.].
[7] Strachey (1956).

broad zone of compromise, since increased labour costs can be passed on through prices and since there is a convergence of the middle-range demands of both sides on the state—demands that aim at increasing productivity, qualifying labour power, and improving the social situation of the workers.[8] The monopolistic sector can, as it were, externalize class conflict.

The consequences of this immunization of the original conflict zone are: (*a*) disparate wage developments and/or a sharpening of wage disputes in the public service sector; (*b*) permanent inflation, with corresponding temporary redistribution of income to the disadvantage of unorganized workers and other marginal groups; (*c*) permanent crisis in government finances, together with public poverty (that is, impoverishment of public transportation, education, housing and health care); and (*d*) an inadequate adjustment of disproportional economic developments, sectoral (agriculture) as well as regional (marginal areas).

In the decades since World War II the most advanced capitalist countries have succeeded (the May 1968 events in Paris notwithstanding) in keeping class conflict latent in its decisive areas; in extending the business cycle and transforming periodic phases of capital devaluation into a permanent inflationary crisis with milder business fluctuations; and in broadly filtering the dysfunctional secondary effects of the averted economic crisis and scattering them over quasi-groups (such as consumers, schoolchildren and their parents, transportation users, the sick, the elderly, etc.) or over natural groups with little organization. In this way the social identity of classes breaks down and class consciousness is fragmented. The class compromise that has become part of the structure of advanced capitalism makes (almost) everyone at the same time both a participant and a victim. Of course, with the clearly (and increasingly) unequal distribution of wealth and power, it is important to distinguish between those belonging more to one than the other category.

The question whether, and if so how, the class structure and the principle of organization that developed in liberal capitalism have been altered through class compromise cannot be examined from the point of view of what role the principle of scarcity and the mechanism of money play at the level of the social system. For the monetization of landed property and of labour, and the

[8] O'Connor (1973).

'progressive monetization of use values and areas of life that were heretofore closed off to the money form', do not indicate conclusively that exchange has remained the dominant medium of control over social relations. Politically-advanced claims to use values shed the commodity form, even if they are met with monetary rewards. What is decisive for class structure is whether the real income of the dependent worker is still based on an exchange relation, or whether production and appropriation of surplus value are limited and modified by relations of political power instead of depending on the market mechanism alone.

ERNEST MANDEL

Late Capitalism and Imperialism*

Three variant types of relation between the bourgeois nation state and the international centralization of capital must be distinguished here. The international centralization of capital may be accompanied by the international extension of the power of *one single state*. This tendency was already observable in the First World War, and in the course of the Second World War and its aftermath it found spectacular expression in the world-wide political and military hegemony of US imperialism. It basically corresponds to the first of the two major forms of the international centralization of capital: decisive control over an increasing share of the international apparatus of production by the owners of a single national class of capitalists, with foreign capitalists participating at most as *junior partners*. The increasing international power of a single imperialist state is congruent with the growing international supremacy of a single national group of capital owners in the total field of international capital.

The international centralization of capital may also be accompanied by a gradual dismantling of the power of the various bourgeois national states and the rise of a *new, federal*, supranational *bourgeois state power*. This variant, which seems at least possible, if not even probable, for the West European EEC

* From Ernest Mandel, *Late Capitalism* (London: New Left Books, 1975), pp. 326–9. Reprinted by permission of New Left Books.

area, corresponds to the second major form of the international centralization of capital: the international fusion of capital without the predominance of any particular group of national capitalists. Just as no kind of hegemony is tolerated in these really multinational companies, the state form corresponding to this form of capital cannot in the long-run involve the supremacy of a single bourgeois nation state over others, nor a loose confederation of sovereign nation states. It must rather take the form of a supranational federal state characterized by the transfer of crucial sovereign rights.

It would certainly be a grave mistake to treat purely economic forces as absolute in this respect and to divorce them from the overall historical context. It is not only the immediate economic interests of capital-owners—or of the decisive group of capitalists in each phase of the capitalist mode of production—that the bourgeois state functions to safeguard. To perform this role effectively, in fact, it must also extend its activity to all the spheres of the superstructure, a task which presents great difficulties if it is undertaken without careful consideration of the national and cultural peculiarities of each particular nationality.[1] In the late capitalist epoch, the direct or indirect economic functions of the bourgeois state apparatus are pushed so far into the foreground— by the constraint to gain increasing control over all the phases of the processes of production and reproduction—that under certain conditions monopoly capital may undoubtedly consider a certain division of labour between a supra-national federal state and cultural activity by nation states a lesser evil. It should not be forgotten that in the United States, for example, all questions concerning education, religion, and culture have—ever since the foundation of the Union—remained in the hands of the individual states rather than those of the Federal Government. Moreover, regulation of educational and cultural questions in various languages is by no means impossible (witness the cantonal system of the Swiss Federation).

The overwhelming compulsion towards the creation of a supra-national imperialist state in Western Europe—if the international centralization of capital were in fact to take the

[1] The particular emphasis on this non-economic superstructural factor explains why the French Gaullists hold fast to the axiom of European 'small states' and why they resist the 'supranationality' represented by 'soulless Eurocrats'.

predominant form of capital fusion on a European level without the hegemony of any one national bourgeois class—springs precisely from the immediate economic function of the State in late capitalism. Economic programming within the nation state is incompatible in the long-run with multinational fusion of capital.[2] The first will either force back the second, especially in periods of crisis or recession, or the second will have to create an international form of programming congruent with itself.[3]

The choice between these two alternatives will ultimately come to a head over the issue of anti-cyclical economic policy, for a successful struggle against crises and recessions, in harmony with the interests of multinational companies, cannot be conducted on a national level; it can only be international. Since the instruments of anti-cyclical policy consist of monetary, credit, budgetary, tax, and tariff devices, such a policy must ultimately have at its disposal a uniform international currency, and a uniform international line on credit, budgeting and taxes (a common international trade policy is already a reality in the EEC). But it is impossible in the long-run to have a common currency, a common budget, a common system of taxes and a common public works programme[4] without a federal government with sovereignty in matters of taxation and finance, and with an executive power of repression to enforce its authority—in other words, without a common State. It should also be said that big multinational companies also create a multinational capital market which in any case makes the survival of national

---

[2] This is the reason why we have for several years expressed the view that the EEC is not yet finally 'irreversible' and could still fall victim to a severe general recession.

[3] The latter must be understood in a twofold sense: in the first place, quantitatively—in other words, a type of economic programming which could set in motion great enough masses of anti-cyclical resources by the State to cope with conjunctural difficulties of realization and sales experienced by huge companies such as Siemans, Phillips, FIAT, or ICI; in the second place, qualitatively—in other words, a type of economic programming capable of quelling particularist regional interests to the wider benefit of the largest multinational companies.

[4] Scitovsky pointed out as early as 1958 that structural and unemployment crises would inevitably result from the creation of the EEC, and argued that a common employment and infrastructural policy (or a policy of public works) would therefore in the long run prove equally inevitable in the EEC: *Economic Theory and Western European Integration*, London 1967, pp. 97–8.

currencies, national credit policies and national budgets and taxes more and more problematic.[5]

The third possible variant of the relationship between the international centralization of capital and the development of the late capitalist State is that of a relative *indifference* of the former to the latter. The example of big British, Canadian and some Dutch companies, in particular, is often cited in this connection.[6] It is customary to emphasize that these companies have internationalized their activities to such an extent, and produce and realize surplus-value in so many countries, that they have become largely indifferent to the development of the economic and social conjuncture of their mother country.[7]

Without denying the existence of this variant, we may, however, regard it as basically no more than an intermediate between the two main variants outlined above. For on closer analysis we must distinguish between two different cases in the operations of these 'state-indifferent' companies. There is the case in which they operate in countries where national state power is itself so weak that it offers no resistance to the quest for additional profits by expatriate concerns: this is ultimately only true of, say, semi-colonial countries controlled by British capital. Or there is the case in which they operate in countries where the national state power that intervenes in the economy is independent of them. With further intensification of international competition and the centralization of capital, the countries in the first group will tend to become increasingly liable to use what State power they have at their disposal to defend their own interests from possible competitors. In the countries of the second group, however, the position of 'state-indifferent' companies is liable to

[5] Several authors have already pointed out the role played by multinational companies in thwarting national attempts to stabilize interest and currency rates in recent years. See for instance, Levinson, op. cit., pp. 36–7, 70–1; Tugendhat, op. cit., p. 161. We shall deal with this problem in chs. 13 and 14.

[6] Robert Rowthorn (with the collaboration of Stephen Hymer), *International Big Business 1957–1967*, Cambridge, 1971, pp. 62–3, 74.

[7] See among others Robert Rowthorn, 'Imperialism: Unity or Rivalry?', in *New Left Review*, No. 69 (Sept.–Oct. 1971), pp. 46–7. Robin Murray, 'Internationalization of Capital and the Nation State', in *New Left Review*, No. 67 (May–June 1971), pp. 104–8, acknowledges the contradiction and concludes that late capitalism is becoming increasingly unstable, without noting that the big companies *must* therefore seek a State power adequate to their needs.

become increasingly threatened by those corporations that enjoy the real support of the local State apparatus. It is then only a question of time before such companies abandon their attitude of indifference to the State and seek to dominate either their home State or the local State within whose frontiers the bulk of their operations takes place. If they fail, these once 'indifferent' companies may have to pay a high price for having underestimated the role of the State in the epoch of late capitalism; they will ultimately fall to their competitors.[8]

FURTHER READING
*

Baran, Paul A., and Sweezy, Paul M. *Monopoly Capital* (1966)
Brewer, A. *Marxist Theories of Imperialism: A Critical Survey* (1980)
Frank, A. G. *Capitalism and Underdevelopment in Latin America* (1969)
Hilferding, Rudolf *Finance Capital* (1981)

---

[8] In the recession year of 1974, even very large corporations like British Leyland or Citroen could only be saved from bankruptcy by massive subsidies from their national governments. But these are corporations which are just below the limit of what national states in Western Europe can still sustain. Multinationals like Phillips, ICI, Siemens, Fiat, or Rhône-Poulenc would need subventions on such a scale, in case of serious financial crisis, that no single national government in capitalist Europe could provide them.

PART VIII

Socialist Society

PRESENTATION

*

One of the least developed parts of Marxist sociology is undeniably that which concerns the structure of socialist societies. Marx himself, in his postscript to the second German edition of *Capital* (1873), referred ironically to the reproach addressed to him by the *Revue Positiviste* (Paris) for confining himself to a critical analysis of existing conditions 'instead of writing recipes (Comtist ones?) for the cookshops of the future'. Nevertheless, it is clear that Marx did have a general conception of what the 'emancipation of humanity' would entail in the way of social organization, in the future 'society of associated producers'; a conception which was expressed in largely philosophical terms in his early writings and was later given a more precise sociological content in various passages of *Capital*, in his study of the Paris Commune, and in his *Critique of the Gotha Programme*.

At all events, Marx's reasonable reluctance, in his own day, to expound a comprehensive and detailed blueprint for a future society in the style (say) of Fourier, becomes irrelevant once there are actually existing socialist societies. Such societies are now a proper, and important, object of sociological study. More than that, from a Marxist standpoint they should be particularly open to critical sociological inquiry, as societies which, in principle, plan and construct their social life ('make their own history') in a rational and uncoerced way, free from class domination. In practice, however, the situation has been very different. During the first few years after the Russian Revolution there was a great deal of critical discussion and cultural innovation, but an orthodox doctrine then came to prevail, enforced during the Stalinist period by terror; and since 1956, in spite of some relaxation of official control, it has remained extremely difficult, in the USSR and the other socialist societies of Eastern Europe (as well as elsewhere), to develop and express the kind of critical sociological thought which would provide the framework for a more profound analysis of the structure and history of these societies. This situation is reflected in the fact that critical studies of how the socialist societies actually function have been produced largely by Western sociologists (Marxist and non-Marxist).

Nevertheless, the limited and uncertain movement towards a liberalization of intellectual life during the past twenty-five years has permitted the gradual emergence of a more critical sociology; notably in Yugoslavia, which broke at an early stage with Stalinist doctrine and practice. Hence it has become possible to explore and debate somewhat more freely—though not yet, in most cases, to investigate by means of substantial empirical research—questions concerning the nature of political power in a 'classless society', the role of bureaucracy, the difficulties of centralized economic planning, the significance of workers' self-management as an embodiment of Marx's conception of a 'society of associated producers', and its operation in practice, the possibility and implications of a 'socialist market economy' as a means of improving the efficiency of both production and distribution.

The following texts illustrate Marx's own conception of socialism, as well as the kind of sociological analysis which is now being directed, from within, upon the problems of existing socialist societies. If this type of critical analysis is extended it will both reflect the influence of reforming movements in the state socialist societies, and contribute to their transformation; but it will also, just as significantly, have an effect upon the character and scope of the movement towards socialism in the present day capitalist welfare states.

KARL MARX

*

*Private Property and Communism**

Communism is the *positive* abolition of *private property*, of *human self-alienation*, and thus, the real *appropriation* of *human* nature, through and for man. It is therefore the return of man himself as a *social*, that is, really human, being, a complete and conscious return which assimilates all the wealth of previous development. Communism as a complete naturalism is humanism, and as a complete humanism is naturalism. It is the definitive resolution of the antagonism between man and Nature, and between man and man. It is the true solution of the conflict between existence and essence, between objectification and self-affirmation, between freedom and necessity, between individual and species. It is the solution of the riddle of history and knows itself to be this solution. (. . .)

We have seen how, on the assumption that private property has been positively abolished, man produces man, himself and then other men, how the object which is the direct activity of his personality, is at the same time his existence for other men, and their existence for him. Similarly, the material of labour, and man himself as a subject, are the point of origin as well as the result of this movement (and because there must be this *point of origin*, private property is a historical *necessity*). Therefore, the *social* character is the universal character of the whole movement; as society itself produces *man* as *man*, so it is *produced* by him. Activity and mind are *social* in their content, as well as in their *origin*; they are *social* activity and *social* mind. The *human* significance of Nature only exists for *social* man, because only in this case is Nature a *bond* with other *men*, the basis of his existence for others and of their existence for him. Only then is Nature the *basis* of his own *human* existence, and a vital part of human reality. The *natural* existence of man has become his *human* existence and Nature itself has become, for him, human. Thus *society* is the

---

* From Karl Marx, *Economic and Philosophical Manuscripts* (1844). Third manuscript. Translated by Tom Bottomore.

accomplished union of man with Nature, the veritable resurrection of Nature, the realized naturalism of man and the realized humanism of Nature. (...)

Private property has made us so stupid and partial that an object is only *ours* when we have it, when it exists for us as capital or when it is directly eaten, drunk, worn, inhabited, etc., in short, *utilized* in some way. But private property itself only conceives these various forms of possession as *means of life*, and the life for which they serve as means is the *life* of *private property*—labour and creation of capital.

Thus *all* the physical and intellectual senses have been replaced by the simple alienation of *all* these senses; the sense of *having*. The human being had to be reduced to this absolute poverty in order to be able to give birth to all his inner wealth. (On the category of *having* see Hess in *Einundzwanzig Bogen*.[1])

The supersession of private property is, therefore, the complete *emancipation* of all the human qualities and senses. It is such an emancipation because these qualities and senses have become *human*, from the subjective as well as the objective point of view. The eye has become a *human* eye when its *object* has become a *human*, social object, created by man and destined for him. The senses have, therefore, become directly theoreticians in practice. They relate themselves to the thing for the sake of the thing, but the thing itself is an *objective human* relation to itself and to man, and vice versa.[2] Need and enjoyment have thus lost their *egoistic* character and nature has lost its mere *utility* by the fact that its utilization has become *human* utilization. (...)

Since, however, for socialist man the whole of *what is called world history* is nothing but the creation of man by human labour, and the emergence of Nature for man, he therefore has the evident and irrefutable proof of his *self-creation*, of his own *origins*. Once the *essence* of man and of Nature, man as a natural being and Nature as a human reality, has become evident in practical life, in sense experience, the search for an *alien* being, a being outside man and Nature (a search which is an avowal of the unreality of man and Nature) becomes impossible in practice. *Atheism*, as a denial of this unreality, is no longer meaningful, for

---

[1] Moses Hess, in *Einundzwanzig Bogen aus der Schweiz*, (1843), p. 329 [Eds.].

[2] In practice I can only relate myself in a human way to a thing when the thing is related in a human way to man.

atheism is a *denial of God*, and seeks to assert by this denial the *existence of man*. Socialism no longer requires such a roundabout method; it begins from the *theoretical and practical sense perception* of man and Nature as real existences. It is a *positive* human *self-consciousness*, no longer a self-consciousness attained through the negation of religion, just as the real life of man is positive and no longer attained through the negation of private property (*communism*). Communism is the phase of negation of the negation, and is consequently, for the next stage of historical development, a *real* and necessary factor in the emancipation and rehabilitation of man. *Communism* is the necessary form and the active principle of the immediate future, but communism is not itself the aim of human development or the final form of human society.

KARL MARX
*
*Communist Society**

What we have to deal with here is a communist society, not as it has *developed* on its own foundation, but, on the contrary, just as it *emerges* from capitalist society; and which is thus in every respect, economically, morally, and intellectually, still stamped with the birth-marks of the old society from whose womb it emerges. Accordingly, the individual producer receives back from society—after the deductions have been made—exactly what he contributes to it. What he has contributed to it is his individual quantum of labour. For example, the social working day consists of the sum of the individual hours of work; the individual labour-time of the individual producer is the part of the social working day contributed by him, his share in it. He receives a certificate from society that he has furnished such and such an amount of labour (after deducting his labour for the common funds), and with this certificate he draws from the social stock of means of consumption as much as costs the same amount of labour. The same amount of labour which he has given to society in one form he receives back in another.

* From Karl Marx, *Critique of the Gotha Programme* (1875). Translated by Tom Bottomore.

Here obviously the same principle prevails as that which regulates the exchange of commodities, as far as this is exchange of equal values. Content and form are changed, because under the altered conditions no one can give anything except his labour, and because, on the other hand, nothing can pass into the ownership of individuals except individual means of consumption. But, as far as the distribution of the latter among the individual producers is concerned, the same principle prevails as in the exchange of commodity-equivalents: a given amount of labour in one form is exchanged for an equal amount of labour in another form.

Hence, *equal right* here is still in principle—*bourgeois right*, although principle and practice are no longer at loggerheads, whereas the exchange of equivalents in commodity exchange only exists *on the average* and not in the individual case.

In spite of this advance, *equal right* is still burdened with bourgeois limitations. The right of the producers is *proportional* to the labour they supply; the equality consists in the fact that measurement is made with an *equal standard*, labour.

But one man is superior to another physically or mentally and so supplies more labour in the same time, or can labour for a longer time; and labour, to serve as a measure, must be defined by its duration or intensity, otherwise it ceases to be a standard of measurement. The *equal* right is an unequal right for unequal labour. It recognizes no class differences, because everyone is only a worker like everyone else; but it tacitly recognizes unequal individual endowment, and thus natural privileges in respect of productive capacity. *It is, therefore, in its content, a right of inequality, like every right.* Right by its very nature can consist only in the application of an equal standard; but unequal individuals (and they would not be different individuals if they were not unequal) can only be assessed by an equal standard in so far as they are regarded from a single aspect, from one particular side only, as for instance, in the present case, they are regarded *only as workers*, and nothing more is seen in them, everything else being ignored. Further, one worker is married, another not; one has more children than another, and so on. Thus, with an equal performance of labour, and hence an equal share in the social consumption fund, one individual will in fact receive more than another, one will be richer than another, and so on. To avoid all these defects, right instead of being equal would have to be unequal.

But these defects are inevitable in the first phase of communist society as it is when it has just emerged after prolonged birth-pangs from capitalist society. Right can never be higher than the economic structure of society and the cultural development conditioned by it.

In a higher phase of communist society, when the enslaving subordination of the individual to the division of labour, and with it the antithesis between mental and physical labour, has vanished; when labour is no longer merely a means of life but has become life's principal need; when the productive forces have also increased with the all-round development of the individual, and all the springs of co-operative wealth flow more abundantly—only then will it be possible completely to transcend the narrow outlook of bourgeois right and only then will society be able to inscribe on its banners: From each according to his ability, to each according to his needs! (. . .)

'Present-day society' is capitalist society, which exists in all civilized countries, more or less free from medieval adjuncts, more or less modified by the special historical development of each country, and more or less developed. On the other hand, the 'present-day State' changes with a country's frontier. It is different in the Prusso-German Empire from what it is in Switzerland, it is different in England from what it is in the United States. '*The* present-day State' is, therefore, a fiction.

Nevertheless, the different States of the different civilized countries, in spite of their manifold diversity of form, all have this in common, that they are based on modern bourgeois society, only one more or less capitalistically developed. They have, therefore, also certain essential features in common. In this sense it is possible to speak of the 'present-day State', in contrast with the future, in which its present root, bourgeois society, will have died away.

The question then arises: What changes will the State undergo in communist society? In other words, what social functions will remain there which are analogous to the present functions of the State? This question can only be answered scientifically, and one does not get a flea-hop nearer to the problem by any number of juxtapositions of the word 'people' with the word 'State'.

Between capitalist and communist society lies the period of the revolutionary transformation of the one into the other. There corresponds to this also a political transition period in which the State can be nothing but *the revolutionary dictatorship of the proletariat*.

KARL MARX

Wealth in a Socialist Perspective*

We never find in antiquity an inquiry into which form of landed property, etc., is the most productive, creates maximum wealth. Wealth does not appear as the aim of production, even though Cato may well investigate which way of cultivating fields brings the highest return, or Brutus may lend his money at the most favourable rates of interest. The inquiry is always into what kind of property creates the best citizens. Wealth appears as an end in itself only among the few trading peoples—monopolists of the carrying trade—who live in the pores of the ancient world like the Jews in medieval society. Now, wealth is on one side a thing, realized in things, in material products, which the human being confronts as a subject; on the other side, as value, it is merely command over alien labour, not for the purpose of domination, but for private enjoyment, etc. In all its forms it appears in the shape of a thing, whether this is an object or a relation mediated by the object, which is external and accidental to the individual.

Thus the old view, in which the human being (however narrowly defined in national, religious, or political terms) always appears as the aim of production, seems very exalted in contrast with the modern world, where production appears as the aim of humanity and wealth as the aim of production. In fact, however, when the limited bourgeois form is stripped away, what is wealth other than the universality of individual needs, capacities, pleasures, productive powers, etc., created by universal exchange? The full development of human mastery over the forces of nature, those of so-called 'Nature' as well as of humanity's own nature? The absolute working out of his creative abilities without any presupposition other than the antecedent historical development, which makes this totality of development, i.e. the development of all human powers as such, an end in itself (not measured by some *predetermined* yardstick) in which humanity does not reproduce itself in one specific form, but produces its totality, strives not to remain what it has become, but is in the absolute movement of becoming? In bourgeois political economy—and in

---

* From Karl Marx, *Grundrisse der Kritik der politischen Ökonomie* (1857–8), 1953 edn., pp. 387–8. Translated by Tom Bottomore.

the epoch of production to which it corresponds—this complete elaboration of human potentialities appears as a total depletion, this universal objectification as total alienation, and the destruction of all fixed, one-sided aims as sacrifice of the human end in itself to a wholly external end. From one aspect, therefore, the childlike world of antiquity seems loftier; and on the other hand, it really is so if we look for an accomplished shape, for form and established limitation. It provides satisfaction from a limited point of view, whereas the modern world gives no satisfaction; or, where it appears satisfied with itself, is *base* and *vulgar*.

SVETOZAR STOJANOVIĆ

*

Socialism and Democracy*

The historical task of socialism is, first, to *universalize* democracy, introducing it to society as a whole, and second, to make it *as direct as possible*.[1] Complete universalization is possible only if the most direct possible mode of decision-making develops, and vice versa. Social ownership is the *conditio sine qua non* of such democratization. But the dependence in this relation is not only unidirectional: ownership can be truly social only insofar as socialist democracy, in the sense in which we have outlined it thus far, exists. It is justly emphasized in Yugoslavia that social self-government, based upon workers' self-management, constitutes

* From Svetozar Stojanović, *Between Ideals and Reality: A Critique of Socialism and its Future*. Translated by Gerson S. Sher (New York: OUP, 1973), pp. 98–101, 115–17, 130–4. Copyright © 1973 by Oxford University Press, Inc. Reprinted by permission of Oxford University Press, New York.

---

[1] Still, we cannot assert (as is occasionally done in Yugoslavia) that socialist democracy, in contrast to other types of democracy, *is direct*. From this error, another has followed: the identification of socialist democracy with complete decentralization. Although it undoubtedly presupposes a great degree of direct and decentralized decision-making, socialist democracy cannot exclude mediation and centralization. But the character of such mediation and centralization must be essentially transformed: All the higher organs of social self-government must grow organically out of the basic forms of the people's self-rule.

the backbone of this kind of democracy. After all, democracy is essentially nothing but social self-rule. Therefore, we can say that in history the degree of democracy has varied directly with the degree that the citizens have actually ruled society.

True socialism ought to lead to an explosion of democracy. Martin Buber foresaw this as a 'structurally rich society' which will make possible free association and dissociation, as well as massive social experimentation.[2] In fact, only social self-government can be a truly open and pluralistic system. Although it has made a significant contribution to the treasury of freedom, bourgeois democracy has been limited mainly to the pluralism of political parties, which has to this day largely remained 'democracy without the people.'[3]

It does not at all follow, however, that democracy in capitalism, in contrast to democracy in socialism, is purely formal. There are far too many such vulgar critiques of bourgeois democracy and tasteless apologies for socialist democracy in Marxist thought. In fact, democracy will always be an uncompleted effort to increase the reality and decrease the formality of the people's chances to participate in decision-making. In socialism, too, it is much easier to create new democratic forms and institutions than to enable people to make use of them effectively, en masse, actively and as directly as possible.

In contrast to other oligarchies, the Stalinists feel the need to conceal their pessimistic attitude about the ability of the people to govern society by a purely futuristic optimism. According to the Stalinists, the people will be ripe for this task only in the communist future. Socialism which would attempt to introduce the self-rule of the people would allegedly be merely utopian and scientifically unfounded. Socialist society, which is ruled by professional politicians and managers, should only create the preconditions for future communist self-rule—material and cultural wealth, an abundance of leisure time, a high level of social consciousness, and so on. The first task is supposedly to develop democratic self-consciousness; only after that, according to the Stalinists, can social self-government be introduced.

However, democratic consciousness does not arise before, but only in the course of, democratic praxis. It is accurate to say that

[2] See his *Paths in Utopia* (New York, 1950).
[3] See Maurice Duverger, *La Démocratie sans le peuple* (Paris, 1967).

the idea of self-government *presumes* universal responsibility, although there are many people who do not act responsibly. But it is just as true that people cannot act responsibly before they take some responsibility upon themselves. If the citizens do not take responsibility upon themselves, however, the revolutionary élite will begin to abuse the concept of the 'transitional period' to antidemocratic and conservative ends.

Stalinists intimately believe that oligarchy is more efficient than the self-rule of the people. However, the concept of social efficiency understood in the oligarchic sense incorporates only the speed with which established goals are achieved, rather than their social justifiability or acceptability. Since social efficiency must be taken to denote the most rapid satisfaction possible of truly *social* interests and the positive recognition and correction of mistakes in decision-making, then in the long run no form of government can be more socially and morally efficient than social self-government.

Of course, in order to neutralize man's tendency to escape from democracy, it is not enough to put people into the process of decision-making and simply abandon them to their own devices. In Yugoslavia we have neglected both in theory and in practice an educational programme of preparation for qualified self-government. Yet it remains quite true that the ability to perceive the entire ensemble of possible decisions, the relationships between short-term and long-term interests, as well as the ability to compare alternatives, are all dependent upon the level of one's education.

However, if people do not desire to participate, if they do not have democratic habits and are not educated to the task of self-government, then self-government will always remain more formal than real and effective. In this case the ruling party, teams of experts, and various informal and other oligarchic groups will find it easy to manipulate people who vainly believe that they themselves really rule. In order for government to be truly self-government, it must be the fruit of a thorough acquaintance with the matter at hand and of an autonomous attitude on the part of all the citizens. (. . .)

A political movement which does not strive to create real possibilities for the introduction of workers' self-management is not a *workers'* movement in the full sense of the term. The criterion for this should be the extent to which the movement takes advantage of *all* opportunities to achieve this goal, not only after it has come to power, but while in opposition as well.

Many contemporary social democratic parties, seized by the vision of the 'welfare state,' view workers' self-management as a utopia which ought to be abandoned at the earliest possible opportunity. While they do not deny that workers' self-management would introduce democracy into the economy, they claim that it cannot function in this sphere because the professionals *must* be given completely free rein. No wonder, then, that other people reason in the same way; George Lichtheim, for example, asserts that workers' self-management is a 'syndicalist utopia' and does not even think that this might need to be demonstrated.[4]

Whenever it has abandoned the programme of workers' self-management, the communist movement has carefully tried to conceal the fact. At one point this may be accomplished by means of the maximalist policy which characterizes the struggle for workers' self-management in capitalism as reformist deception. When the movement comes to power, workers' self-management is postponed to the unforeseeable future. Stalinists profess unmeasured 'love' for the working class, but in fact they have no faith in it. Stalinism's futuristic ideological mask fell from its face the moment it proclaimed the beginning of the construction of communism and once again displaced workers' self-management to the future.

Elements of capitalism arose even before the bourgeois revolution, which merely cleared the political obstacles from the path of its development. In contrast to the bourgeois revolution, the socialist October Revolution was presented only after its victory with the task of creating the initial elements of the new society. In the previous society there were elements of the first signs of statization, but not of true socialization.

This is why the undeveloped state of the theory of socialist society had such serious consequences. I am not suggesting, of course, that the revolution finally resulted in statist degeneration primarily because of the weaknesses of Marxist theory. Nevertheless, it should be emphasized that Marx wrote only sporadically about self-government. In *State and Revolution*, Lenin went into raptures over Marx's analyses of the Paris Commune,

---

[4] See *Marxism in Modern France* (New York, 1966), p. 144.

but after the assumption of power he gradually leaned toward the position that self-management ought to be postponed until the country was industrialized. Thus a gap gradually developed between the working class's revolutionary creativity and revolutionary theory. The working class spontaneously created its own organs of self-management in the revolution. But instead of taking precisely these organs as the model of the new form of social organization, theory increasingly oriented itself toward the state. Of course, practice changed as well. The role of the soviets gradually decreased. The unity of theory and practice was finally re-established, but only after both had lost their revolutionary character.

Today we know quite a bit about statism. Socialism, however, still seeks its content. Yet there is one thing which is beyond all doubt: socialism cannot be constructed without workers' self-management. The fact is that in nearly all the revolutionary stirrings of our century the working class has spontaneously attempted to introduce its self-management. Even after communists have come to power, the working class has manifested the same tendency—in Poland and Hungary in 1956, and twelve years later in Czechoslovakia.

The workers will become the socially dominant class primarily by means of their self-management. In order to exist, however, workers' self-management must be seen as the core of social self-government. The working class cannot preserve and improve upon its self-management if the other areas of social life are abandoned to the monopoly of the state apparatus, because the latter will soon begin to strangle workers' democracy itself. This is how we should interpret Marx's idea that the proletariat cannot emancipate itself unless it simultaneously emancipates the entire society. Only through the development of workers' self-management as the basis for social self-government will the proletariat fully become a class for itself. Without this, however, socialist revolutions cannot be social but only political ones, and they gradually degenerate into statism.

Marxists still have serious work to do in constructing a theory of social self-government. This holds for Yugoslav Marxists as well, although they have, of course, accomplished the most in this respect. The fundamental contributions should not be expected solely from philosophy or any of the social sciences taken separately. A synthetic approach is indispensable. In this respect it would be well for Marxists to make free use of the theoretical fund of other socialist currents—for example syndicalism and Fabianism—which have actually cultivated the idea of workers' self-management more than have Marxists themselves. It should also be emphasized at this point that a successful theory of social self-government can be only a part of a broader theory of socialist community. (. . .)

After the lively discussions and practical vacillations that culminated in the New Economic Policy (NEP), the Bolshevik Party decided upon extreme centralization and a distributive economic model. This choice was one of the fateful factors in establishing the basis for primitive-politocratic statism. It was once again confirmed that the mode of production determines the nature of the entire social system. Among the Bolsheviks the view was victorious that rapid industrialization imperatively demands complete state centralization of accumulation. They came to the conviction that socialism is irreconcilable with market economy.

Without going into the question of whether this belief was based upon the authentic Marx or not, one thing should be emphasized as indisputable: in order for social products to lose their character as commodities, it is necessary, according to Marx, that they exist in abundance. But here was a society of *scarcity* which wanted to abolish market economy. The result was statism, not socialism. The state planners and the distributive apparatus became alienated from society and began to take the lion's share of social production for itself. This economic and sociopolitical model later became a paradigm for an entire group of countries.

The choice of the non-market economic model after the revolution was also undoubtedly motivated by the desire to avoid the irrationality of anarchic and wasteful market competition by means of strict state planning. However, statist subjectivism and voluntarism have produced results no less irrational.

Without a commodity-money economy today there cannot be any rapid progress toward material abundance. There is still no better way in which to determine the *material* needs of the population and on this basis to produce real 'use values'. The otherwise noble wish to eliminate the mediation of 'exchange values' at the very beginning of socialism has resulted in production for stockpiles rather than for use.

The market can diagnose and cure many 'infantile disorders' and shatter many economic myths constructed by statism. It can reveal, for example, that beneath full employment there is only pseudo-employment, or that an extraordinary rate of economic growth can to a certain degree mean accumulation of unused stock. Behind the facade of economic dynamism the market can reveal the harmful investment errors of the politocracy, which has tried to erect monuments to itself in the form of so-called political factories. In a word, the market can unmask parasitism and introduce effective selection and stimulation.

In the absence of the pressure of real market competition, self-managing collectives in Yugoslavia used to react rather indifferently to the use-values of their products. Moreover, their economic interests were not such as to induce them toward mutual integration. Self-government is threatened not only by statism, but also by a utopian picture of human nature, on the basis of which people naïvely expect that self-managing groups will produce rationally at any given moment, without any competition whatsoever. In a system without competition, solidarity is shattered by its opposite—parasitism.

In order to preserve a given society's socialist character, however, the market must be placed within the framework of serious planning, regulation, and co-ordination. Otherwise, economics and morality will continue to react *antagonistically* upon each other, economics pulling in the direction of egoism, and morality in the direction of solidarity. Without rational control of economic tendencies by the associated producers, socialism in Marx's sense is out of the question. However, neither in theory nor in practice has an economic model yet been discovered which might synthesize self-management, the market, and planning.

John Maynard Keynes was among the first to perceive that bourgeois economic thought was the victim of two major errors: (1) that the economy can successfully regulate itself by means of the market mechanism; and (2) that an uncontrolled market ensures the maximum utilization of economic potentials. Thus, Yugoslav anarcho-liberals who *glorify* the market in socialism are provincially repeating the mistakes of pre-Keynesian bourgeois economics. In place of the demolished myth of statist de-alienation, they are constructing a myth of de-alienation through the uncontrolled market. But the idea of socialist community is as irreconcilable with servitude to the blind forces of the market as it is with statist alienation.

So long as it exists, the market will try to impose itself over society as the supreme regulator and criterion of human relations so that it may thereby restore the economic basis of bourgeois society. It is generally recognized that the market reacts mainly to the existing level of demand and that it also creates artificial and even harmful demands. It thus comes into conflict with the mission of socialist community, which seeks to humanize existing need and develop new, human needs. This notwithstanding, the 'socialist' anarcho-liberals persistently attack Marxists who assert that humanism and the money-commodity economy often come into conflict.

This conflict is quite obvious in culture, although the market could reveal some parasites here as well. Yugoslav experience only confirms what we already know from capitalism—that individual groups (self-governing, now) can use the market to encourage the most uncultured of needs and make quite a bit of money in the process. It is indisputable that capitalist civilization still dictates our structure of needs and consumption to a considerable degree. This has been the source of much confusion over the question of the proper criteria of socialism. Even the very concept of the standard of living is increasingly reduced to a material standard, rather than being seen as a human standard.

In Yugoslavia there are theorists and practitioners who advocate simply transferring market principles to the field of culture. But socialism, if understood in a Marxist sense, should orient itself toward the *gradual* elimination of cultural values from the list of commodities. Statism has replaced the rule of *homo economicus*, so characteristic of bourgeois society, with the domination of *homo politicus*. In the name of socialism the anarcho-liberals want to transcend *homo politicus* only to rehabilitate *homo economicus*. Actually, both are equally pernicious for socialist culture.

The quality of socialist society depends to a large degree upon the manner in which the consumer is educated. In our country quite a bit has been said about the association of producers, but all too little about the association of consumers. In contrast to group-particularistic self-government, in a system of *integral* social self-government it is possible for needs to be democratically hierarchicalized, for priorities to be established and for means to exist to educate and satisfy the most human of needs. A theory of such needs, which for its part would assume the further development of Marxist anthropology and axiology, would make further elaboration of the theory of socialist community possible.

GEORGE KONRAD and IVAN SZELENYI
*
Intellectuals, Workers, and Political Power*

The social structure of early socialism is organized in keeping with the principle of rational redistribution. In line with the rational principle on which its economy is based, we regard this as a class structure, and indeed a dichotomous one. At one pole is an evolving class of intellectuals who occupy the position of redistributors, at the other a working class which produces the social surplus but has no right of disposition over it. This dichotomous model of a class structure is not sufficient for purposes of classifying everyone in the society (just as the dichotomy of capitalist and proletarian is not in itself sufficient for purposes of assigning a status to every single person in capitalist society); an ever larger fraction of the population must be assigned to the intermediate strata.

We consider as belonging to the middle strata those who neither possess redistributive power nor engage in direct productive labour themselves; those who occupy low-level supervisory positions in which they communicate redistributive decisions to the workers but which carry no orienting or cross-contextual functions and so do not involve intellectual knowledge; and those who partake of society's goods on the basis of some other legitimating principle, such as the few remaining small owners (who even still employ wage workers occasionally), small craftsmen and shopkeepers, and private peasants. (. . .)

In modern redistributive systems it is the rationality of the

* From George Konrád and Ivan Szelényi, *The Intellectuals on the Road to Class Power* (Brighton: Harvester Press, 1979), pp. 145, 147–8, 173–5, 222–4, 229. Copyright © 1979 by Harvester Press Ltd. and Harcourt, Brace, Jovanovich, Inc. Reprinted by permission of Harvester Press and Harcourt, Brace, Jovanovich, Inc.

redistributors' activity which legitimates their authority. The power of the social group which exercises a monopoly over the redistributive process has a class character; it is the power of the intellectual class even though the functions of central redistribution in the strict sense are carried out not by the intelligentsia as a whole but by a narrower segment of it—the state and party bureaucracy, which we shall call the ruling or governing élite, since the criteria for admission to it are those of a status group rather than a class.

Modern redistribution replaces the decisions of the market with official administrative decisions which, in the aggregate, call into being a bureaucratic organization that tends to become highly centralized and monolithic, and to encompass the whole of society. Important political and economic decisions are made on the upper levels of the élite bureaucracy, and these upper-level positions must be occupied by intellectual officials. Not every intellectual takes part in making important redistributive decisions, but every major decision is made by the intelligentsia of office.

It is a further important characteristic of rational-redistributive society that it does away with the separation of economic, political, and cultural power, a division of powers which is taken for granted, almost like a constitutional principle, in societies with market economies. Thus the powers of the redistributors are far more than just economic powers. Redistribution creates a whole model of civilization, in which three partners of equal importance participate: the stratum of economists and technocrats, which actually carries out the work of central redistribution; the administrative and political bureaucracy, which guarantees the undisturbed functioning of the redistributive process, by police measures if necessary; and finally the ideological, scientific, and artistic intelligentsia, which produces, perpetuates, and disseminates the culture and ethos of rational redistribution. The ideologue, the policeman, and the technocrat are mutually dependent and are impregnated with one another's logic. (. . .)

According to the ideology of the Communist parties socialism is the social system in which the working class is the ruling class; socialism means the dictatorship of the proletariat. What is there to support this cardinal tenet of all Communist theory? What empirical facts point to the rule of the working class under socialism? Do workers receive the highest pay? Hardly; the average intellectual's earnings are considerably higher than the average worker's, and the differential between the maximum an intellectual can earn and the maximum worker's wage is as great if not greater than in the market economies. Are the workers able to take greater advantage of state-subsidized benefits over and above wages? No; white collar people live in larger and more comfortable dwellings in pleasanter neighbourhoods, and they have a far better chance than workers do of getting apartments in buildings being constructed with the aid of state subsidies. Even the right to settle in towns is a class privilege: it is easier for intellectuals to get permission to settle in the cities with their superior services, and so they can live relatively close to their places of work, while a good part of the working class—in some countries as much as half—is obliged to commute to work from the ill-served villages where they live. The children of the intelligentsia go on to university-level studies in far higher proportion; even earlier, they gain admittance to better schools more easily than do workers' children. Only intellectuals and their dependants are admitted to the special hospitals and clinics which provide outstanding care for ranking state and party officials. (. . .) As a result of all these perquisites the differentials between the living standards of workers and intellectuals are far greater than their officials earnings alone would indicate. (. . .)

If the leading role of the working class cannot be demonstrated from any tangible aspect of material life, such as income or living standard, are the workers compensated perhaps by a greater voice in the making of decisions? Does the leading role of the working class mean that actual workers have a little more to say than others do in deciding essential economic issues or even just the shop-floor technical questions that arise in their factories? In fact the worker, for all his alleged 'leading role', has just as little to say in the high- or low-level decisions of his enterprise as the worker in a capitalist plant. He has no voice in deciding whether operations will be expanded or cut back, what will be produced, what kind of equipment he will use and what direction (if any) technical development will take, whether he will work for piece rates or receive an hourly wage, how performance will be measured and production norms calculated, how workers' wages will evolve relative to the profitability of the enterprise, or how the authority structure of the plant, from managing director to shop foreman, will operate. (. . .)

Workers, as workers, cannot intervene in the policies of their enterprises because in the socialist countries there are no real workers' organizations. The Communist parties, after coming to power, quickly dissolved or transformed every organization in which only workers participated, from workers' councils, factory committees, and trade unions to workers' singing societies, theatrical groups, and sports clubs. In the Socialist countries only corporative organizations exist, and workers belong to them only in company with the administrative personnel and technical intelligentsia of their enterprise or branch. (. . .) It is hardly a coincidence that whenever a political upheaval culminates in a workers' rebellion—which always comes just as unexpectedly and just as terrifyingly for the opposition intelligentsia as for the ruling élite, as in Budapest in 1956 or Gdánsk in 1970—the first order of business for the workers is to form their own, non-corporative organizations: workers' councils or soviets. Nor was it any accident that after the defeat of the Hungarian uprising of 1956 the workers' councils, formed without the participation of intellectuals, were the first to be dissolved; and even though the retrospective 'white books' containing the official ideological evaluation of the event speak only of the leaders of the intellectual opposition, the great majority of the death sentences imposed (apart from those on armed rebels) fell on members of the workers' councils, who in general took no part in the armed uprising. (. . .)

Rational—redistributive society, as we have said, can best be described as a dichotomous class structure in which the classical antagonism of capitalist and proletarian is replaced by a new one between an intellectual class being formed around the position of the redistributors, and a working class deprived of any right to participate in redistribution. Under rational redistribution, however, social conflicts do not appear in the form of open class conflict, since in the absence of organic class intelligentsias class interests cannot be articulated openly, and no class can develop a clear consciousness of its strategic goals. Indeed, can we even say that the classes have different interests if these do not find expression in distinct ideologies? (. . .)

If there are two classes under rational redistribution, let us see if we can sketch in broad outline the conflicts of interest between them. Clashes of interest in society, particularly between classes, can be described in terms of conflicts between opposing principles of legitimation. To challenge the power of the class above it a subordinated class must shake the legitimacy of its authority in its own eyes and before society as a whole. Alongside (or rather beneath) the legitimating principle of the ruling class the legitimating principle of the oppressed is always to be found; though it may not pervade the culture of the age it is always empirically discoverable in the consciousness of the labouring classes. Under rational redistribution it is above all technical knowledge, intellectual knowledge, which legitimizes the right to dispose over the surplus product. That is what justifies the superior position of the redistributors and provides the ideological basis for the formation of the intelligentsia into a class. Those, on the other hand, who have been deprived of any power of disposition over the surplus product—the very people who work together to produce it—can appeal to only one alternative principle of legitimation in order to challenge the power of the intellectual class, and that is the legitimating principle of possession of labour power. It is the essence of that principle that those who produce the surplus product should dispose over it, not those who claim that they know better how it should be distributed. (. . .)

It is in the interests of the working class that a larger share of the national income should turn up in the pay envelopes of the workers, while a smaller share goes to the state budget for redistribution. They have an interest in seeing the state spend more of the existing budget on consumer subsidies and less on weapons, and in having more state investment funds allocated to infrastructural development, especially of a sort that meets the communal needs of the population, rather than to new productive ventures of dubious value. The interests of the working class demand, further, that government subsidy of consumption should be guided by social welfare considerations, so that those products and services which workers and low-income people in general need should receive the most support. Thus the interests of the workers are diametrically opposed to those of the redistributors not only in respect to wages and other shop-floor issues, but also in matters of macro-economic planning as well.

FURTHER READING

*

Bahro, Rudolf *The Alternative in Eastern Europe* (1978)
Broekmeyer, M. J. (ed.) *Yugoslav Workers' Self-Management* (1970)
Brus, Wlodzimierz *The Economics and Politics of Socialism* (1973)
Hegedüs, András, Heller, Agnes, Markus, Maria, and Vajda, Mihaly *The Humanisation of Socialism* (1976)
Lane, David *The Socialist Industrial State* (1976)
Parekh, Bhikhu (ed.) *The Concept of Socialism* (1975)

BIBLIOGRAPHY

*

The Bibliography lists all the major works referred to in the Introduction and texts, except (1) those works from which excerpts have been taken, where details of publication are provided in a footnote, and (2) the works of Marx and Engels, where details are given either in the text or in footnotes, since the contributors refer to various editions.

Abercrombie, Nicholas; Hill, Stephen; and Turner, Bryan S., *The Dominant Ideology Thesis*. London: Allen & Unwin, 1980.
Adler, Max, *Der soziologische Sinn der Lehre von Karl Marx*. Leipzig: C. L. Hirschfeld, 1914.
Anderson, Perry, *Passages from Antiquity to Feudalism*. London: New Left Books, 1974.
Anon., *The Source and the Remedy of the National Difficulties*. London, 1821.
Arato, Andrew; and Breines, Paul, *The Young Lukács and the Origins of Western Marxism*. London: Pluto Press, 1979.
Aron, Raymond, *Progress and Disillusion*. London: Pall Mall Press, 1968.
Avineri, Shlomo, *The Social and Political Thought of Karl Marx*. Cambridge: CUP, 1968.
Bahro, Rudolf, *The Alternative in Eastern Europe*. London: New Left Books, 1978.
Baran, Paul A.; and Sweezy, Paul M., *Monopoly Capital*. New York: Monthly Review Press, 1966.
Bauer, Otto *Die Nationalitätenfrage und die Sozialdemokratie*. 1907. 2nd enlarged edn. Vienna: Wiener Volksbuchhandlung, 1924.
Baxandall, Lee; and Morawski, Stefan (eds.) *Marx and Engels on Literature and Art*. New York: I. G. Editions, 1974.
Berlin, Isaiah *Karl Marx*. Oxford: OUP, 1963.
Bernal, J. D. *The Social Function of Science*. London: Routledge, 1939.
Bernard, Claude *Introduction à l'étude de la médecine expérimentale*. Paris: J. B. Baillière, 1865.
—— *Leçons sur les phénomènes de la vie communs aux animaux et aux végétaux*. Paris: J. B. Baillière, 1878–9.
Bernstein, Eduard *Evolutionary Socialism*. 1899. English trans. New York: Schocken Books, 1961.
Blumberg, Paul, *Industrial Democracy: The Sociology of Participation*. London: Constable, 1968.
Bottomore, Tom, *Marxist Sociology*. London: Macmillan, 1975.
—— *Sociology as Social Criticism*. London: Allen & Unwin, 1975.

—— (ed.), *Modern Interpretations of Marx*. Oxford: Basil Blackwell, 1981.
—— and Goode, Patrick (eds.) *Austro-Marxism*. Oxford: OUP, 1978.
—— and Rubel, Maximilien (eds.), *Karl Marx: Selected Writings in Sociology and Social Philosophy*. Harmondsworth: Penguin Books, 1963.
Braverman, H., *Labor and Monopoly Capital*. New York: Monthly Review Press, 1974.
Brewer, Anthony, *Marxist Theories of Imperialism: A Critical Survey*. London: Routledge & Kegan Paul, 1980.
Broekmeyer, M. J. (ed.), *Yugoslav Workers' Self-Management*. Dordrecht: D. Reidel Publishing Co., 1970.
Brus, Wlodzimierz, *The Economics and Politics of Socialism*. London: Routledge & Kegan Paul, 1973.
Brym, Robert, *Intellectuals and Politics*. London: Allen & Unwin, 1979.
Buber, Martin, *Paths in Utopia*. London: Routledge & Kegan Paul, 1949.
Bukharin, Nikolai, *Historical Materialism: A System of Sociology*. 1921. English trans. New York: International Publishers, 1925.
—— *Imperialism and the Accumulation of Capital*. 1926. English trans. edited by K. J. Tarbuck. New York: Monthly Review Press, 1972.
Carrillo, Santiago, *'Eurocommunism' and the State*. London: Lawrence & Wishart, 1977.
Childe, V. Gordon, *Man Makes Himself*. 3rd edn. London: Watts, 1956.
Cohen, G. A., *Karl Marx's Theory of History: A Defence*. Oxford: OUP, 1978.
Cohen, Y., *Man in Adaptation*. New York: Aldine Publishing Co., 1968.
Crosser, P. K., *State Capitalism in the Economy of the United States*. New York: Bookman Associates, 1960.
Cummins, Ian, *Marx, Engels and National Movements*. London: Croom Helm, 1980.
Dahrendorf, Ralf, *Class and Class Conflict in an Industrial Society*. London: Routledge & Kegan Paul, 1959.
De Vore, Irven, and Lee, Richard B., *Man the Hunter*. Chicago: University of Chicago Press, 1966.
Döbert, Rainer, *Systemtheorie und die Entwicklung religiöser Deutungssysteme*. Frankfurt am Main: Suhrkamp, 1973.
Durkheim, Émile, *The Division of Labour in Society*. 1893. English trans. New York: Macmillan, 1933.
—— *The Rules of Sociological Method*. 1895. English trans. New York: The Free Press, 1964.
Duverger, Maurice, *La Démocratie sans le peuple*. Paris: Éditions du Seuil, 1967.
Easton, L., and Guddat, K. (eds.), *Writings of the Young Marx on Philosophy and Society*. New York: Doubleday, 1967.

Ferguson, Adam, *An Essay on the History of Civil Society*. 1767. New edn. edited with an introduction by Duncan Forbes. Edinburgh: Edinburgh University Press, 1966.
Finley, M. I., *Ancient Slavery and Modern Ideology*. London: Chatto & Windus, 1980.
Fischer, Ernst, *The Necessity of Art: A Marxist Approach*. 1959. English trans. Harmondsworth: Penguin Books, 1963.
Frank, A. G., *Capitalism and Underdevelopment in Latin America*. Rev. edn. New York: Monthly Review Press, 1969.
Friedmann, Milton, *Capitalism and Freedom*. Chicago: University of Chicago Press, 1969.
Galbraith, J. K., *The New Industrial State*. Boston: Houghton Mifflin, 1967.
Gay, Peter, *The Dilemma of Democratic Socialism*. New York: Columbia University Press, 1952.
Giddens, Anthony, *Capitalism and Modern Social Theory*. Cambridge: CUP, 1971.
Girard, René, *Mensonge romantique et vérité romanesque*. Paris: Grasset, 1961.
Godelier, Maurice, 'Structure and Contradiction in *Capital*', in Robin Blackburn (ed.), *Ideology in Social Science*. London: Fontana/Collins, 1972.
Goldmann, Lucien, *The Hidden God*. 1956. English trans. London: Routledge & Kegan Paul, 1964.
—— *Recherches dialectiques*. Paris: Gallimard, 1959.
—— *Marxisme et sciences humaines*. Paris: Gallimard, 1970.
—— *Cultural Creation*. Oxford: Basil Blackwell, 1977.
—— *Method in the Sociology of Literature*. Oxford: Basil Blackwell, 1981.
Goldthorpe, John H., Lockwood, David, Bechhofer, Frank, and Platt, Jennifer, *The Affluent Worker in the Class Structure*. Cambridge: CUP, 1969.
Graubard, S. R., *A New Europe?* Boston: Houghton Mifflin, 1964.
Gurvitch, Georges, *Dialectique et sociologie*. Paris: Flammarion, 1962.
Habermas, Jürgen, *Toward a Rational Society*. Boston: Beacon Press, 1970.
—— *Knowledge and Human Interests*. London: Heinemann, 1972.
—— *Legitimation Crisis*. London: Heinemann, 1976.
—— and Luhmann, N. *Theorie der Gesellschaft oder Sozialtechnologie*. Frankfurt am Main: Suhrkamp, 1971.
Hardach, Gerd, and Karras, Dieter, *A Short History of Socialist Economic Thought*. London: Edward Arnold, 1978.
Haupt, Georges, Lowy, Michael, and Weill, Claudie, *Les Marxistes et la question nationale, 1848–1914: Études et textes*. Paris: François Maspero, 1974.

Hayek, F. A. von, *Law, Legislation and Liberty*. 3 vols. London: Routledge & Kegan Paul, 1973–8.
Hegel, G. W. F., *Principles of the Philosophy of Right*. 1821. English trans. Oxford: OUP, 1942.
Held, David, *Introduction to Critical Theory*. London: Hutchinson, 1980.
Herwegh, Georg (ed.), *Einundzwanzig Bogen aus der Schweiz*. Glarus, 1843.
Herzfeld, Levi, *Handelsgeschichte der Juden des Altertums*. Braunschweig: J. H. Meyer, 1879.
Hilferding, Rudolf, *Finance Capital*. 1910. English trans. London: Routledge & Kegan Paul, 1981.
—— *Das historische Problem*. 1941. English trans. of part in Bottomore (1981).
Hindess, Barry; and Hirst, P. Q., *Pre-Capitalist Modes of Production*. London: Routledge & Kegan Paul, 1975.
Hussain, Athar, and Tribe, Keith, *Marxism and the Agrarian Question*. London: Macmillan, 1981.
Jentsch, Karl, *Drei Spaziergänge ins klassische Altertum*. Leipzig: F. W. Grunow, 1900.
Kolakowski, Leszek, and Hampshire, Stuart (eds.), *The Socialist Idea: A Reappraisal*. London: Weidenfeld & Nicolson, 1974.
Lane, David, *The Socialist Industrial State*. London: Allen & Unwin, 1976.
Larrain, Jorge, *The Concept of Ideology*. London: Hutchinson, 1979.
Lefebvre, Henri, *The Sociology of Marx*. New York: Random House, 1968.
Levinson, Charles, *The Multinational Pharmaceutical Industry*. Geneva: International Federation of Chemical and General Workers' Union, 1973.
Lichtheim, George, *Marxism in Modern France*. New York: Columbia University Press, 1966.
Löwith, Karl, *Max Weber and Karl Marx*. London: Allen & Unwin, 1982.
Lukács, Georg, *The Theory of the Novel*. 1916. English trans. London: Merlin Press, 1971.
Luxemburg, Rosa, *The Accumulation of Capital*. 1913. English trans. London: Routledge & Kegan Paul, 1951.
—— *The Russian Revolution*. 1922. English trans. Ann Arbor: University of Michigan Press, 1961.
Mallet, Serge, *The New Working Class*. Nottingham: Spokesman Books, 1975.
Mann, Michael, *Consciousness and Action among the Western Working Class*. London: Macmillan, 1973.
Marcuse, Herbert, *One-Dimensional Man: Studies in the Ideology of Advanced Industrial Society*. London: Routledge & Kegan Paul, 1964.

Marquard, Odo, *Schwierigkeiten mit der Geschichtsphilosophie*. Frankfurt am Main: Suhrkamp, 1973.
Mehring, Franz, *Die Lessing-Legende*. Frankfurt am Main: Ullstein, 1972.
Melman, S., *Pentagon Capitalism: The Political Economy of War*. New York: McGraw-Hill, 1970.
Miliband, Ralph, *The State in Capitalist Society*. London: Weidenfeld & Nicolson, 1969.
Moore, Barrington, *Social Origins of Dictatorship and Democracy*. London: Allen Lane, 1967.
Mosca, Gaetano, *Elementi di scienza politica*. 2nd revised and enlarged edn. 1923. English version under the title *The Ruling Class*. New York: McGraw-Hill, 1939.
Murra, John, 'The Economic Organization of the Inca State'. Unpublished Ph.D. thesis, University of Chicago, 1956.
Nairn, Tom, *The Break-Up of Britain: Crisis and Neo-Nationalism*. London: New Left Books, 1977.
Neilson, N., *Customary Rents*. Oxford: OUP, 1910.
O'Connor, James, *The Fiscal Crisis of the State*. New York: St. Martin's Press, 1973.
Ossowski, Stanislaw, *Class Structure in the Social Consciousness*. London: Routledge & Kegan Paul, 1963.
Parekh, Bhikhu (ed.), *The Concept of Socialism*. London: Croom Helm, 1975.
Perrier, Edmond, *La Philosophie zoologique avant Darwin*. Paris: Félix Alcan, 1884.
Pokrovsky, M. N., *Brief History of Russia*. 2nd edn. Orono, Me.: University Prints, 1968.
Poulantzas, Nicos, *Political Power and Social Classes*. London: New Left Books, 1973.
—— *Classes in Contemporary Capitalism*. London: New Left Books, 1975.
Prawer, S. S., *Karl Marx and World Literature*. Oxford: OUP, 1976.
Pribicevic, Branko, *The Shop Stewards' Movement and Workers' Control, 1910–22*. Oxford: Basil Blackwell, 1959.
Reagan, M. D., *The Managed Economy*. New York: OUP, 1963.
Renner, Karl, *The Institutions of Private Law and their Social Functions*. 1904. English trans. London: Routledge & Kegan Paul, 1949.
Rose, H., and Rose, S. (eds.), *The Political Economy of Science*. London: Macmillan, 1976.
Rowthorn, Robert, *International Big Business, 1957–1967*. Cambridge: CUP, 1971.
Sahlins, M., *Tribesmen*. Englewood Cliffs, N. J.: Prentice-Hall, 1968.
Saint-Simon, Henri Comte de, *Le Nouveau Christianisme*. 1825. English trans. in F. H. M. Markham (ed.), *Henri Comte de Saint-Simon: Selected Writings*. Oxford: Basil Blackwell, 1952.

Schumpeter, J. A., *Capitalism, Socialism and Democracy*. 5th edn. London: Allen & Unwin, 1976.
Scitovsky, Tibor, *Economic Theory and Western European Integration*. London: Allen & Unwin, 1958.
Shaw, William H., *Marx's Theory of History*. London: Hutchinson, 1978.
Shonfield, A., *Modern Capitalism*. Oxford: OUP, 1965.
Smith, Adam, *An Inquiry into the Nature and Causes of the Wealth of Nations*. London, 1776.
Sombart, Werner, *Why is there no Socialism in the United States?* 1906. English trans. London: Macmillan, 1976.
Soustelle, Jacques, *La Vie quotidienne des Aztèques à la veille de la conquête espagnole*. Paris: Hachette, 1955.
Stalin, Joseph, *Dialectical and Historical Materialism*. New York: International Publishers, 1972.
Strachey, John, *Contemporary Capitalism*. London: Gollancz, 1956.
Sweezy, Paul M., *Modern Capitalism and Other Essays*. New York: Monthly Review Press, 1972.
—— et al. *The Transition from Feudalism to Capitalism*. New edn., with an Introduction by Rodney Hilton. London: New Left Books, 1976.
Taine, Hyppolite, *Le Gouvernement révolutionnaire*. A volume of *Les Origines de la France contemporaine*. Paris: Hachette, 1876–94.
Touraine, Alain, *The Post-Industrial Society*. New York: Random House, 1971.
Tugendhat, C., *The Multinationals*. London: Eyre & Spottiswoode, 1973.
Turner, Bryan S., *Marx and the End of Orientalism*. London: Allen & Unwin, 1978.
Valéry, Paul, *Pièces sur l'art*. Paris: Maurice Darantière, 1931.
Varga, E., *The Great Crisis and its Political Consequences: Economics and Politics 1928–1934*. London: Modern Books, 1935.
Vinogradoff, P., *Villeinage in England*. Oxford: OUP, 1891.
Weber, Max, *Economy and Society*. 1922. English trans. 3 vols. New York: Bedminster Press, 1968.
Weidenbaum, M. I., *The Modern Public Sector*. New York: Basic Books, 1969.

INDEX

absolutism 138
Adler, Max 6, 15
Adorno, Theodor W. 161
aesthetics 161, 188
agriculture 52, 75, 240
alienation 42
Allende, Salvador 226
Althusser, Louis 5, 6
Appianus 70
aristocracy 94, 134, 169, 170
Aristophanes 70
army 127
  people's 195
art 166, 167
  Greek 167
Asiatic mode of production 56, 74–86, 215
Association pour l'Étude des Questions Sociales 26
Athens 69, 127
Austro-Marxism 119, 153
authority, political 139, 140

Balzac, Honoré de 186, 187
Baran, Paul 235
base and superstructure 5, 47
Bauer, Bruno 41
Bauer, Otto 147
Benjamin, Walter 161
Berlin, Isaiah 225
Bernard, Claude 28
Bernstein, Eduard 150, 152, 254fn.
Bismarck, Otto von 129, 245
Bolsheviks 149, 284
Bolshevism 148
Bonapartism 96, 128
Bourbons 96
bourgeoisie 25, 35, 94, 95
  rural 175
  urban 175
Brazil 227–9
  Brazilian miracle 227
Buber, Martin 280
Bukharin, Nikolai 101, 235
bureaucracy 288
  Soviet 115
Byzantine Empire 53

Caesar, Julius 197
Cam, Helen 81
capital
  accumulation of 207, 210
  centralization and concentration of 207, 243, 264, 265
  finance 254
  international 230
capitalism
  advanced 258–64
  competitive 258
  international 224, 248
  late 144, 145, 247, 264–8
  liberal 258, 260, 263
  managed 150
  organized 150, 236, 258–64
  state 236
  state monopoly 236
  transition from feudalism to 84, 85
Carrillo, Santiago 119
Carthage 72
Cassel, Professor 251
caste 103, 107
Cavaignac, General 96
Chaplin, Charles 191
chemistry 247
Childe, V. Gordon 64
China 40, 74, 75
Chile 226
civil society 19, 35, 36, 122–4, 176
class (*see also* bourgeoisie, middle class, working class) 92, 99–102
  artisan 73, 74
  consciousness 107–12
  dominant 128, 141
  domination 132
  service class 133–6, 153
  struggle 33, 221
clergy, Catholic 105
Code Napoléon 98
commodity 85
  economy 253
  production 210
communism 273–8
community 56
  cultural 197
  village 76, 77, 85

Index

Comte, Auguste 34, 38, 39
conflict theorists 140, 142
conquest 53
consciousness 180, 181
  bourgeois 188
  class 107–12
  collective 181
  democratic 280
  self 20
contradictions 110
contrat social 21
corporation *see* joint-stock company
crisis
  of capitalism 242, 243
critical theory *see* Frankfurt School
Croce, Benedetto 171
culture 161–3
  national 197
Czechoslovakia 283

democracy 139, 196–8, 279–87
  bourgeois 119
  formal 147, 261
dependency 222–30
despotism, oriental 75, 76, 78, 79
determinism 137
  economic 47
dialectic
  of the social process 33
dialectical materialism 42
dictatorship of the proletariat 277
Diodorus Siculus 71
division of labour 23, 51, 52, 55, 122–4, 210, 217
Durkheim, Emile 1, 23–9, 151–2
  Rules of Sociological Method 23

ecclesiastics 170
economics (*see also* political economy) 31
economies
  central and peripheral 224
  underdeveloped 224
economy
  market 284
  organized 247–53
educational system 174, 195
EEC (European Economic Community) 265
empiricism 61, 63
embourgeoisement 151fn.
Enfantin, Barthélemy-Prosper 41
Engels, Friedrich
  letters: to J. Bloch 6, 47; W. Borgius 6, 47; F. Mehring 161; C. Schmidt 6; F. Tönnies 34fn.
  Ludwig Feuerbach and the Outcome of Classical German Philosophy 37
  Origin of the Family, Private Property and the State 26
  Socialism, Utopian and Scientific 37, 242–7
Estate
  Fourth 107
  Third 103
estate system 55
Eurocommunism 119
evolution
  regressions in 217
  social 216
expropriation
  of expropriators 208

Fabianism 284
Fascism 139, 192, 193
feudalism 37, 81–6, 169
Feuerbach, Ludwig 31, 41
Fichte, Johann Gottlieb 42
film 192
Fourier, Charles 243, 244, 271
Frankfurt School 9, 161
free competition 245, 247, 249, 251
Futurism 193

gens 76, 77, 126
gentile associations 126
Gericke, H. 218fn.
Girard, René 178, 181, 183
Godelier, Maurice 6fn., 8, 9
Goldman, Lucien 7, 162
  The Hidden God 42
  'Y-a-t-il une sociologie marxiste?' 42
government 252
Gramsci, Antonio 5, 9
  Prison Notebooks 135fn.
Guevara, Che 226
guilds 74
guild-masters 106
Guizot, François 49
Gurvitch, Georges 8

Hauptman, Gerhard
  The Weavers 196
Hegel, G. W. F. 35, 36, 41, 49, 126, 211–12
  Phenomenology of Spirit 41
  Principles of the Philosophy of Right 36
Hegelians, left 41
hegemony 5
Heidegger, Martin 188

Hellenic world   72
Hess, Moses   274
Hilferding, Rudolf
   Finance Capital   9fn., 120, 235, 255fn.
history
   of the species   214
Hitler, Adolf   135
Hobsbawm, Eric   203
homology   180
Homo Economicus   286
Homo Politicus   286
Hungary 1956   283

idealism   18
   historical   28
ideology   35, 161–3
   bourgeois   261
   petty bourgeois   40
imperialism   255, 264–8
   U.S.   264
import substitution   229
Independent Social Democratic Party (USPD)   148
India   74, 75
individualization   210
integration theorists   140, 142
intellectuals   168–77, 287–91
intelligentsia   289
Italy   53, 175

Jentsch, Karl   67–8
joint-stock company   245, 246
journeymen   55, 66
Junkers   128

Kafka, Franz   185fn.
Kautsky, Karl   9, 255
Keynes, J. M.   285
Kierkegaard, Sören   42
kinship   127

labour
   market   250
   movement   196
   productivity of   197
   serf   55
   services   83
   social   239
   surplus   240
   time   241
Lassalle, Ferdinand   106
latifundia   73
Latin America   222–9
Leach, Edmund   63

legitimation   260, 291
Leibniz, Gottfried Wilhelm   42
Lenin, V. I.
   'A Great Beginning'   100–1
   Imperialism   255fn.
Lévi-Strauss, Claude   6, 65
Lewis, Herbert   65
Licinius   53
Lipset, S. M.   151
literature   161–2, 178–88
lithography   190
Loria, Achille   177
Luhmann, N.   216fn.
Lukács, Georg   181
lumpenproletariat   53, 70, 74
Luxemburg, Rosa   9fn., 110
   Die Akkumulation des Kapitals   255fn.

Maine, H. S.   80, 82
Marcuse, Herbert   152, 162
Marinetti, Emilio   193
mark, the   126
materialism
   dialectical   40
   historical   212–18
Mehring, Franz
   Die Lessing-Legende   161
Mexico   227
Middle Ages   37, 54, 66, 74, 82, 83, 93–5
middle class   108, 152, 153
Mikhailovsky, N. K.   4
monarchy
   absolute   93, 128
   legitimist   124
   July   124
monopoly   245
Morgan, L. H.   61, 80, 104
Mosca, Gaetano   169fn.
Mussolini, Benito   135, 148
mythology   167

Napoleon Bonaparte   97
nation   196–7
national interest   222
nationalism   10
national underdevelopment   222
nation state   10
   bourgeois   265
nature   241, 274, 278
Nazism   139
Neilson, N.   84
neo-evolutionism   61–3, 217
NEP (New Economic Policy)   284
Niebuhr, Reinhold   126fn.

nobility
    feudal 105
noblesse
    de robe 104, 171
    d'épée 104
norms 25
nouveau roman 185
novel, the 178–88

Odyssey, the 68
Ordine Nuovo 173
oriental despotism (*see also* Asiatic mode of production) 75, 76, 78, 79

Pan-Africanism 155
Pareto, Vilfredo 169fn.
Paris Commune 120
parties 150, 252
    Communist 288
peasantry 54, 70, 96, 169–70, 195
Persia 74
Peru 226
petty bourgeoisie (*see also* middle class) 93
philosophy
    German 17
photography 190
Piaget, Jean 7
Picasso, Pablo 191
planning, global 259
Plato 42
plebeians 53
plebs 124
pluralism 144
Pokrovsky, M. N. 82
political economy 41, 143
    bourgeois 211
population 22
Positivism 34
Poulantzas, Nicos 7, 137fn.
praxis 33, 43
prices 250
production
    capitalist 238, 239
    commodity 210
    forces of 49, 242
    in general 22
    modes of Part II *passim*, 214, 215
    planned 249
    relations of 49, 214
professional associations 107
progress
    Marx's concept of 208–12
proletariat *see* working class

property
    asiatic form 56
    communal 52, 56
    estates 54
    feudal 54
    forms of 51–55
    landed 92
    private 53, 273, 274
    small-holding 98
    State 52
    tribal 52
Proudhon, P. J. 39, 41
psychology 24

rationality, economic 223
realism 42
redistribution 288, 290
reification 110, 180, 183
Reichstag, the 146
relativism 42
religion 104
Renner, Karl 162
republic
    democratic 128, 129
    parliamentary 124
Revue Positiviste 271
revolution
    bourgeois 20, 187
    Communist 220, 221
    French 1789 109, 220
    industrial 65
    neolithic 65
    proletarian 150
    Russian 1917 109, 271
    social 218–21
    technological 116
    total 220
Rheinische Zeitung 35
Ricardo, David 20, 37
right
    bourgeois 276
    equal 276
Robbe-Grillet, Alain 185fn.
Robinsonades 21
Roman Empire 54
Rome 53, 57, 127
Roosevelt, Franklin D. 135
Rousseau, Jean-Jacques 21
rules
    of instrumental action 213
    of strategic action 213

Saint-Simon, Henri Comte de 35, 39, 121

Sarraute, Nathalie 185fn.
Schmoller, Gustav 82
Schumpeter, Joseph A. 150, 154
science
  natural 24, 240
self-management
  workers' 156, 279
serfdom 55, 57, 85
Shakespeare, William 166
slavery 53, 57, 66–74, 127, 128
  domestic 71
Smith, Adam 20, 36
social democracy 95, 147, 156, 250
social formation Part II *passim*
socialism 23, 249, 251, 272–5, 279–87
  and democracy 279–87
socialist societies Part VIII *passim*
social process 33
socialization 30–3
  economic 31
  of labour 207
  process 32
society
  capitalist *see* capitalism
  Communist *see* communism
  feudal *see* feudalism
  gentile 125
  socialist *see* socialism
  tribal 58–66
Sombart, Werner 151
Sorel, Georges 15
species subject 215
speculation 18
Spencer, Herbert 23
Stalinism 280, 281
State
  absolutist 138
  authoritarian 146
  Bismarckian 138
  Bonapartist 129
  debts 127
  democratic 250
  enterprise 229
  German philosophy of the 252
  intervention 143, 251
  relative autonomy of the 120, 132, 138
  socialist 120
  totalitarian 120
  welfare 145
statism 284, 285
status 103–7
  groups 103–5
Stone Age 68

stratum 101, 102
  middle 287
  middle class strata 152
structuralism 6–10
  Marxist 7
structures 9
  economic 215
  interest 215
Struve, P. 82
superstructure 24, 49, 50, 188
surplus
  production 77
  value 77
surrealism 192
Sweezy, Paul M. 235
syndicalism 283

Taine, Hyppolite
  Le Gouvernement révolutionnaire 25
taxes 127
Taylor, Frederick 172
Ten Hours Bill 94, 253
Third World 205
trade unions 116
trusts 245
Turks 53

utilitarianism 27
United States of America 151, 225
USSR 40

Valéry, Paul 190
value
  surplus 240
  use 179
Varga, Eugene 235
Venezuela 229
Vinogradoff, Paul 84

wage labour 66, 69
  and capital 57
wages 251
war 58
  civil 147
wealth 304, 305
Weber, Max 2, 48
workers' councils 112, 156, 290
working class 92, 93, 149–56

Xenophon 70

Yugoslavia 156, 281, 285, 286